Aligning the Organisation to the Vision

MSc in Global Marketing Practice

Aligning the Organisation to the Vision

First edition September 2010

ISBN: 9780 7517 8414 5
Ebook: 9780 7517 8425 1

British Library Cataloguing-in-Publication Data
A catalogue record for this book has been
applied for from the British Library

Published by

BPP Learning Media Ltd
Aldine House, Aldine Place
London W12 8AA

www.bpp.com/learningmedia

Printed in Great Britain

We are grateful to the GMN Advisory Council
member Ian Derbyshire for Module Advisor
feedback.

Author:
Darrell Kofkin

A note about copyright

Contents

Congratulations

on choosing to study towards MSc Global Marketing Practice.

You have made the right decision because you have at your fingertips a wealth of support to help guide you through your studies with BPP Learning Media, Ashcroft International Business School and Global Marketing Network.

GMN's CEO Darrell Kofkin summarises the key value of The Global Marketer Programme as:

"The programme is designed to prepare you thoroughly for a rewarding career in marketing and ensure you can make a profitable contribution to business anywhere in the world."

For the first time, you can achieve an academic award at Masters level from one of the UK's largest and most progressive universities, while at the same time ensuring you meet all the academic criteria required for acceptance as a Professional Member of Global Marketing Network.

It is our intention that at this final level of this innovative Masters programme, MSc Global Marketing Practice, you are able to develop the skills you require to perform effectively from day one.

This Study Text is designed to work in conjunction with the following elements of the overall programme:

- A Student Handbook designed to give you all the information you need to achieve success

- A personal E-Tutor to guide, encourage and develop your thinking and provide guidance for your assignment

- Access to The Global Marketer virtual learning environment providing you with a week-by-week plan of study for each module, with access to articles, case studies, additional learning activities and a global discussion forum connecting you with other programme participants

- Access to the Anglia Ruskin University digital library, providing access to an extensive range of online databases, world leading journals and E-books

- Access to the GMN online network with all the latest news about approved conferences, workshops and events that you can attend to support your continuing development

Each chapter of the Study Text is designed to refer to each Study Session within the programme.

Integrating Global Marketing Network Philosophies

Each module has at least one GMN Advisory Council or Faculty member acting as Module Advisor. Module Advisors contribute to the development of the curriculum and study materials.

The GMN Module Advisors for Aligning the Organisation to the Vision are Ian Derbyshire, Nigel Culkin and Helena Kim.

Ian is Chairman Global Marketing Network and Managing Director IDA. His business life is all about building businesses and leading change. He has run a thriving management consultancy-to-ventures business since 1994, and has helped his impressive portfolio of clients large and small, national and global, to start businesses, to grow businesses and to sell businesses.

Nigel is Head of Enterprise & Entrepreneurial Development at the University of Hertfordshire and Associate Dean of the Business School.

Helena is an Associate Director of Coaching and leadership trainer for i2i, an international leadership and development consultancy. She backs up her work with 20 years of professional experience and research.

Additional information and reviews of the Study Materials by Module Advisors can be found on the Global Marketer Programme supporting Virtual Learning Environment.

Using this Study Text

This Study Text includes features designed specifically to make learning effective and efficient.

- Each chapter begins with an **introduction** to set the chapter in context.

- Chapter **learning outcomes** summarise what you are expected to be able to achieve through reading the chapter. The content covered is summarised in a Chapter Roundup which is organised according to these learning outcomes at the end of the chapter.

- Harvard referencing is used throughout with **a full reference list** at the end of each chapter.

Throughout the Study Text, there are also special aids to learning. These are indicated by the following symbols.

Definition

Definitions are provided for key terms.

Activity 1

Activities for you to complete are included throughout the chapters. Often these will ask you to relate theories to your own organisation. Where appropriate activitiy debriefs are included at the end of the chapter.

Global case study

Real life global examples are used to illustrate marketing practice.

 Readily available key text book resources, significant chapters and additional reading are selectively added to the chapters to help develop your further reading.

 Selected online materials which link to the topics covered within the chapter are suggested at appropriate points.

Assessment advice

Points to bear in mind in relation to specific chapter topics when preparing your assignment are inserted at relevant places in the chapter.

GMN viewpoint

GMN experts' opinions, snippets from presentations and papers are included to bring both practitioner and academic cutting edge perspectives into the material.

About this module

The module provides you with a critical and, where necessary, targeted insight into the latest issues relating to creativity and innovation, leadership and change so that they can recommend, develop, implement and measure marketing strategies that bring real and sustainable competitive advantage. Where possible, the emphasis will be placed on your own organisation. It covers three key areas:

1 Creativity and Innovation

The module will critically evaluate practical organisational issues which require the application of lateral thinking and creative techniques. It will investigate the paradox of logic and creativity in the strategic development process. It will critically evaluate the importance of 'intrapreneuring' and how to create a climate of innovation that brings out the entrepreneurial spirit of the organisation. It will investigate ways to achieve a culture of innovation and entrepreneurship within the organisation and align the organisation and its resources with the vision to achieve sustainable competitive advantage.

2 Leadership

The module will investigate ways to improve one's own leadership effectiveness within the organisation, drawing upon a range of classical approaches to leadership including: traits, behavioural, styles, charisma and transformational leadership theories plus contemporary thinking around leadership including: emotional intelligence, psychodynamics, ethical & spiritual leadership and leader-led relationships. Practical approaches that work and are readily grasped/applied at all levels will be included such as 'Situational Leadership', 'Towards the Complete Leader', 'Shaping the Future', together with practical ways to integrate the desired leadership behaviours which tend to be found in successful organisations worldwide within the performance management process in the student's own organisation. The factors that tend to deliver successful teamwork and followership will be discussed in this context.

3 Change Management

The module will critically evaluate practical change projects within the organisation which are "live" or have been completed within the recent past, drawing upon proven models of change management. A straightforward and practical approach to change management and organisational transformation will be described and studied, including how to make change happen, how to make it sustainable and how to get change right through to the customer interface through a proven workshop-driven cross-functional approach involving all stages in the main business process / supply chain.

Module aims

Both the study mode and module content further facilitate the application of, and reflection on, theory in practice. You will be expected to work on an assignment related either to your own organisation or a national/international business organisation.

The module supports the development of relevant employability, promotability and professional skills which enables you to develop an understanding of the organisation.

Learning outcomes for the module are to:

1   Demonstrate an in-depth understanding of the theories, concepts and techniques required to successfully align the organisation to the brand vision to achieve sustainable competitive advantage.

2   Assess the role of creativity and innovation as a way to achieve a culture of innovation and entrepreneurship in the organisation.

3   Apply methods and techniques to improve leadership effectiveness within the organisation in order to integrate the desired leadership behaviours which tend to be found in successful organisations worldwide.

4   Analyse methods to drive change management and organisational transformation within the organisation involving all stakeholders in the main business process and supply chain.

5   Reflect critically upon own learning throughout the module, and articulate the value of this learning to an organisation's and to the student's own professional development.

A note on pronouns

On occasions in this Study Text, 'he' is used for 'he or she', 'him' for 'him or her' and so forth. While we try to avoid this practice it is sometimes necessary for reasons of style. No prejudice or stereotyping accounting to sex is intended or assumed.

Key contacts

Please email globalmarketer@bpp.com for any queries about this Study Text and The Global Marketer Programme.

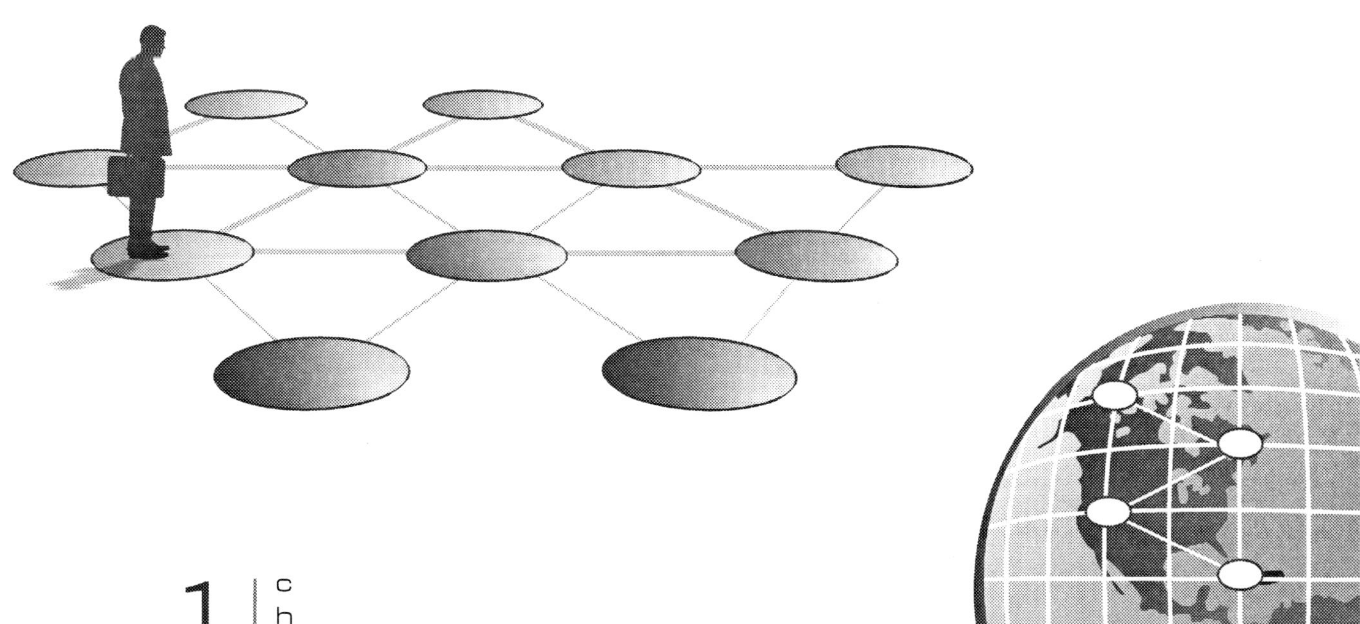

1

chapter

The importance of organisational vision in leading the organisation

The most successful organisations and the most successful leaders have a clear vision and mission. Lack of attention in this area will lead to strategic thinking that is confused, muddled and often misunderstood. Close attention in this area will ensure that leaders are able to clearly communicate their organisation's vision to all of their stakeholders and also ensure that organisational objectives are both coherent and congruent. If people within an organisation do not know what they are expected to achieve and leaders do not set a clear direction, then it will be difficult to create a clear and coherent strategy or course of action for the organisation. Those visions that are most successful are those that are developed with the consideration of the organisation's stakeholders. To be delivered effectively all stakeholders need to feel that they play an important role in its delivery. There needs to be strong leadership, a healthy organisational culture and an organisational structure and methods of operation that encourage creativity and innovation. These aspects are all covered in this module. This chapter sets the scene for this module and explores the importance of an organisational vision, the importance of both the shareholder and the stakeholder and the need for evaluating, developing and maintaining an organisation's core competences. We then look at the theoretical nature of leadership and distinguish between leadership and management.

Contents

1

Chapter learning outcomes

By reading this chapter you will be equipped to:

- Understand the importance of an organisational vision

- Compare and contrast perspectives on organisational purpose

- Evaluate the role of core competences in an organisation's mission

- Evaluate the difference between leadership and management

- Discuss characteristics of leadership and different strategic leadership theories

- Critically suggest how organisational realities impact on leadership

- Consider appropriate alternatives to leadership

1 The importance of organisational vision

GMN viewpoint – Ian Derbyshire, Chairman GMN

The fact that 'Aligning the Organisation to the Vision' is 'only' a 15-credits module is not a reflection of its relative importance in the overall Masters programme. In terms of practical impact, and from a business/practitioner perspective, this is arguably *the* single most important module in the programme – certainly it was conceived as such. Criticisms levelled at professional marketers over the last few years have arguably centred around the connection between 'marketing' and 'what drives an organisation forwards', especially at the most senior levels. This module provides the essentials to help students on this programme redress that perceived historic weakness. It also features, as part of Global Marketing Network's strategic thrust to raise standards in the marketing profession worldwide, in the GMN Continuing Professional Development programmes being set up and run for existing professional marketers.

Leaders and managers must constantly make choices and seek solutions based on an understanding of what their organisation is intended to achieve. They are confronted every day with many different claimants who believe that the organisation exists to serve their interests. Many of us will have to make tough choices and often we are forced to weigh up the interests of different stakeholders and decide which take priority on certain matters.

Where we have a clear understanding of the organisation's vision and mission this can provide strong guidance during processes of strategic thinking, strategy formation and strategic change.

The organisational vision and its guiding mission can act as a guiding light, a fundamental principle against which strategic options can be evaluated and decided upon.

Collins and Portas (1994) measured performance of visionary versus comparison companies and found evidence that companies with a strong vision delivered superior performance. Visionary companies set huge goals and formidable tasks for themselves. It is their willingness to take on and meet daunting challenges that is one of their keys to long-term performance superiority. The researchers concluded that the truly great visionary companies have strong, insular cultures and that the essence of a visionary company comes in the translation of its core ideology into the very fabric of the *organisation*. The vision should seep into everything that the company does.

1.1 Key issues of organisational vision and mission

The concept of organisational vision and mission can be an elusive concept. To many people they are merely lists of principles that while they may have potential PR value, have little if any bearing on the actual business, let alone the process of strategy choice, formulation and implementation.

However, a clear vision and mission can be very concrete and play a critical role in determining strategic decision-making and action, thereby ensuring the organisation and its constituent parts are aligned to its delivery.

1.2 Creating an organisational vision and mission

Definition

Mission comes from the Latin word *'mittere'* which means 'to send'. A mission is a task, duty or purpose that sends some on their way. Hence corporate mission can be understood as the basic drivers sending corporation along its way and consists of the fundamental principles that mobilise and propel the firm in a particular direction.

To understand how a mission impacts on strategy two areas require closer attention:

First, it is necessary to know what types of fundamental principles actually make up a corporate mission.

Second, it needs to be examined what types of roles are played by a corporate mission in the strategy formation process.

The successful direction of the organisation is as a result of the effective interaction between management and the board of directors. The board of directors must be coherent in its view and implementation of the organisational mission.

Its activities, described as 'corporate governance' governs the strategic choices and action of the management of the organisation. The task of corporate governance is to shape, articulate and communicate the fundamental principles that will drive the organisation's activities.

The following example demonstrates how a short and to the point mission can help to drive and inspire the way the organisation operates.

Global case study

The Fed Ex mission under Fred Smith: 'absolutely simply positively overnight'.

Read it again. You do get the point. They deliver. Overnight. Every night. No excuses. It's what we do. A great example of leadership in action, causing marketing hype to meet daily reality in a win:win encounter for the benefit of the customers and, ultimately, the shareholders. Oh, and the staff.

1.2.1 The difference between a mission and a vision

Unless they are well defined internally, there can be a tendency for the vision and mission to become confused or thought of as the same concept.

Definition

De Wit and Meyer (2005) make a distinction between **'Vision and Mission'**. While the corporate mission outlines the fundamental principles guiding strategic choices a *'strategic vision'* outlines the desired future at which the organisation hopes to arrive.

An alternative and very clear view held by GMN Chairman Ian Derbyshire is:

A mission is not a **vision**. The **vision** is at the top of the hierarchy. It isn't to be confused with the day-to-day mission.

The London Underground introduced and communicated widely in 1988-90 – and at every level – a 'vision' and a 'mission' as part of a much larger root and branch business restructuring and process re-engineering (literally when it came to replacing wooden escalator parts and the correcting the absence of effective fire detection and sprinkler systems in tube stations) to help refocus and energise the 24,000 employees in the years following the devastating Kings Cross fire.

Vision: *'To be the best metro in the world'.*

Mission: *'We aim to P.L.E.A.S.E'.*

PLEASE stood for Partner, Leader, Enjoyable, Affordable, Safe, Efficient. Detailed customer research amongst customers and stakeholders alike demonstrated that this is what they required. It all boiled down to something short, simple and memorable. Trite too, to some, but nonetheless, there was a very high percentage recall amongst staff following the internal communication/discussion process. Such clear, focused, aspirational/inspirational leadership and strategic positioning was crucial in winning support for all sorts of huge internal changes – and all done in the context of the fire too – but, in time, valid human aspiration was subjugated to the 'real world reality' of accountants and civil servants and the vision became, with new leaders coming into the most senior roles, *'to create a decently modern metro'*... hardly inspiring, or memorable, or now even much remembered... in time, this too went out with a whimper, together with the hopes and aspirations of very many staff who by then were facing the imposition of what was recognised internally then as – and has since been shown to have been – a wholly unworkable part-privatisation (the PPP). Now all this is very UK-centric, but the general lessons are applicable more widely.

GMN viewpoint – Ian Derbyshire

A 'vision' in this context should describe in a brief one-line, inspiring and future-centred way _where_ the organisation wants to be and/or _what_ it wants to be 'when it gets there' or, you could refer to it as 'when it grows up'. This is absolutely not the same as a 'mission'. And it might be a long way off, but its where the organisation wants to be. Meantime, it focuses day-in-day-out on delivering its mission, which is realisable on pretty much a daily basis.

1.3 Elements of organisation mission

Organisational purpose can be defined as the reason why an organisation exists. The perception that an organisation's management has of its organisation's purpose gives direction to the strategic process and will include the strategic choices and the way it is ultimately implemented.

More often than not a manager's view of the organisation's purpose will be part of a broader set of business principles that steers strategic thinking.

It is this set of principles that forms the base of an organisation's identity and guides its strategic decision-making that is often referred to as the organisational mission.

De Wit and Meyer (2005) identify three other components of a corporate mission as:

- **Organisational beliefs**

    To work in unison a common understanding is needed amongst all organisational stakeholders. The stronger the set of beliefs subscribed to by the stakeholders the easier communication and decision-making will become and the more confident, cohesive and driven the group will be. If people do not share the same beliefs joint decision-making will be protracted and conflict will arise.

- **Organisational values**

  Each person in an organisation can have their own set of values. Yet, when an organisation's stakeholders share a common set of values this can have a strong impact on the strategic direction. Widely held values contribute to a clear sense of organisational identity which will attract some individuals and dispel others. To be of most use and influence such values must be embodied in the organisation's culture.

- **Business definition**

  Most organisations have a clear identity from being active in a particular sector or industry. For such organisations, having a delimiting definition the business they want to be in strongly focuses the direction in which they develop. Their business definition functions as a strong guiding principle and helps to distinguish opportunities from diversions. Here while a clear business definition can focus the organisation's efforts it can also lead to short-sightedness and missing new business developments and opportunities.

The strength of the mission will depend on whether and how effectively the elements, as outlined above, fit together and are mutually reinforcing (Campbell and Yeung, 1991).

When a consistent and compelling mission is formed this can ignite the organisation with a sense of purpose and create an emotional bond as well as a rational reason for emerging the members' work in a manner consistent with the mission.

1.4 Functions of the organisational mission

While an organisation's mission can be articulated by means of a mission statement, in practice, not everything that is called a mission statement meets the above criteria.

Organisations can have a mission statement, even if it has not been explicitly written down. Non-written statements however do risk divergent interpretations and could lead to a dilution of the mission being implemented and acted upon.

There are a number of clear benefits of a clear, coherent and well communicated mission:

- **Direction**

  A mission can point the organisation in a certain direction, by defining the boundaries within which strategic choices and actions must be taken.

- **Legitimisation**

  A mission conveys to all stakeholders inside and outside the organisation what activities it is pursuing and on what basis. Through specifying the business philosophy that will guide the company the chances are increased that stakeholders will accept, support and trust the organisation.

- **Motivation**

  A mission can inspire individuals to work together in a particular way, with the power to motivate people over a prolonged period of time.

- **Strategic change**

  A mission statement, articulated in an inspiring and thought provoking way can enable a change of attitude within an organisation which can lead it in a new direction.

Global case study

An example of a mission statement that engendered change is that of BMW, where the board of directors proclaimed the motto 'enterprise mobility'. From the external viewpoint this does not focus attention on

the product 'cars' but on the mobility as a specific customer need. From the internal standpoint 'enterprise mobility' challenges every single employee and organisational unit in their behaviour.

1.4.1 Practical considerations when creating a mission statement

> **GMN viewpoint – Ian Derbyshire**
>
> There may be no hard and fast rules about how to create an effective mission statement, but I would argue that an effective mission statement does exhibit certain key characteristics The mission is critical and it is that which we are here to do, day-in, day-out. It is not an essay. And an essay is not memorable, nor is it a mission. All other text – for example to explain the organisation's values and beliefs, its strategic positioning etc – has to be seen as subsidiary. This helps to keep the mission statement itself short and to the point which, in practice, is very much more helpful, focused and indeed memorable. 'If it isn't memorable, it isn't any use to anyone.

The key characteristics of an effective mission statement according to GMN Chairman Ian Derbyshire are that it is:

- **Brief** – so that it is a memorable short statement of the task rooted in a clear strategic understanding of the organisation.

- Works at **every level** and within every department of the organisation.

- **Makes sense** and is believed in.

2 Perspectives on organisational purpose

Organisations increasingly face two competing pressures:

(1)   They require a certain amount of economic profitability to survive.

(2)   They need to exhibit a certain amount of social responsibility if they are to retain the trust and support of key stakeholders.

At one end of the debate are those people who argue that corporations exist to serve the purpose of their owners. This point of view is commonly referred to as the 'shareholder value perspective'.

At the other end of the spectrum are those people who argue that corporations should be seen as joint ventures between a number of parties. This is referred to as the **'stakeholder value perspective'**.

The shareholder value perspective	The stakeholder value perspective
Corporations are instruments whose purpose is to create economic value on behalf of those who invest risk-taking capital in the enterprise and should drive the purpose whether they are privately or publicly held.	Corporations are a coalition between various resource suppliers, with the intention of increasing their common wealth. An organisation should be regarded as a joint venture in which all participate to achieve economic success.
Many people making this argument feel that the well-being of the shareholder is best served if the strategy of the company leads to higher share prices and/or higher dividends.	The emphasis shareholders place on share price appreciation and dividends must be balanced against the legitimate demands of the other partners. Demands are both quantitative and qualitative in nature.

The shareholder value perspective	The stakeholder value perspective
Supporters of this perspective state that it is challenging to get management to pursue shareholders' interests where ownership and managerial control have become separated.	Managers must recognise their responsibility towards all constituents since maximising shareholder value to the detriment of other stakeholders is unjust. Managers have a moral obligation to consider the interests and values of all joint venture partners.
The emphasis on profitability does not ignore the need for stakeholder analysis. Paying attention to stakeholders does not mean to serve them as the only duty of the corporation is to maximise shareholder value within the boundaries of what is permissible. There is no moral obligation to treat stakeholders well.	Managing stakeholder demands is an end in itself – it is not only more just but more effective since it is easier to work as a motivated team where all interests are served and there is a greater sense of partnership, ownership and trust.

Activity 1

Talk with other managers in your organisation and identify what behaviours determine their view of the organisation and what they believe should be the purpose of the organisation. Compare and contrast their responses. What does this tell you about the ability to create a mission statement that will be universally accepted?

2.1 Shareholders and stakeholders: A new perspective on corporate governance

If stakeholders are to be engaged in the development of corporate strategy and the organisational mission it is useful to define exactly what a stakeholder is and where the term emanates from. Freeman and Reed (1983) proposed two definitions of stakeholder:

- **The wide sense of stakeholder**

  Any identifiable group who can affect the achievement of an organisation's objectives or who is affected by the achievement of an organisation's objectives.

- **The narrow sense of stakeholder**

  Any identifiable group or individual on which the organisation is dependent for its continued survival. They made the point that organisational strategies need to account for those groups who can affect the achievement of the firm's objectives since this will allow the analysis of all external forces and pressures whether they are friendly or hostile.

2.2 Use of stakeholder concepts in strategy formulation

Definition

Stakeholder Strategy Process is a systematic method for analysing the relative importance of stakeholders, their co-operative potential (how they can help the corporation achieve its objectives) and their competitive threat (how they can prevent the corporation from achieving its objectives).

Stakeholder Audit Process is a systematic method for identifying stakeholders and assessing the effectiveness of current organisational strategies.

By itself, each process analyses the stakeholder environment from the standpoint of the organisational mission and seeks to formulate strategies for meeting stakeholder needs and concerns.

One analytical device depicts an organisation's stakeholders. The first dimension is one of 'interest' and 'stake'. The second dimension is its 'power', which ranges from formal power of shareholders, to the economic power of customers to the political power of special interest groups.

	Formal or voting power	Economic power	Political power
Equity stake	• Shareholders • Directors • Minority interests		
Economic stake		• Customers • Debt holders • Competitors • Unions • Suppliers	• Foreign governments
Influencers			• Consumer groups • Government • Trade associations

2.3 Involving stakeholders in future possibilities

It is often said that if people do not feel involved they will not be committed. Involvement is the key to implementing change and increasing commitment. If we are not involved it is likely that we will resist change. People tend to hold on to old views, old ways, old habits. And old styles and habits are hard to change.

To make or break a commitment takes great commitment. And commitment comes from **involvement** – it acts as a catalyst in the change process.

The downside of involvement is **risk**. Whenever you involve people in the problem you risk losing control. It seems safer and easier not to involve others but simply tell them, to direct them, to advise them. Many young or new managers hesitate to involve people in decision-making for fear of opening up other options or compromising their position.

Eventually, through experience, most managers learn that the effectiveness of their decisions depends on quality and commitment, and that, as stated earlier, commitment comes through involvement. They are then willing to assume the risks and to develop the skills of involving people appropriately.

Activity 2

Develop a stakeholder grid for your organisation and evaluate the importance of each stakeholder in the development and delivery of the organisation's vision.

2.4 Developing an effective mission statement

There are no hard and fast rules about how to create an effective mission statement but here are some issues that you should bear in mind:

- **Ask yourself the right questions**

    To begin, ask yourself these kinds of questions: What business are you in? Why are you in this business? What do you want for yourself, your employees, and your customers?

- **Say it clearly**

  Your mission statement needs to clearly state business goals and objectives. It should explain what the business is, what special niche it inhabits in the marketplace, and how it will make a difference in the lives of customers and clients.

- **Decide what makes your organisation different**

  Never forget that you are pursuing the same customers as your competitors. How is it that you stand out from those other companies? Ask yourself if it is because you do something better, cheaper, or faster than others. Identify any underlying philosophies or values that guide your company.

- **Build your brand**

  Use your mission statement to build your unique brand. Make sure to communicate your business's key value to the customer or client segment that you are serving.

- **Keep it short and sweet**

  Remember: brevity is the soul of wit. A strong mission statement shouldn't ramble on endlessly. Ideally, you should be able to summarise your company mission in a few sentences. Consider it your 'elevator pitch' — you, your board of directors, your managers and your other employees should be able to state your company's mission succinctly in the time it takes to ride an elevator from the ground floor to the top floor.

- **Be honest**

  Make sure that when you read your own mission statement, it reflects what you truly believe. Too much pomp and self-congratulatory language will turn off those who read it, so avoid saying that your company is the 'best' at this and the 'world leader' at that.

- **Make it a joint effort**

  Even if you are the sole proprietor of your business, don't write your mission statement in a vacuum. It's incredibly helpful to get the input of others, both inside and outside the company. Collaborators can help you to better see the strengths and weaknesses of your mission statement. Ask for input, including external stakeholders such as your key clients, suppliers and partners with whom you may have a very close professional relationship.

- **Polish the language**

  See to it that you have several pairs of eyes (ideally belonging to wordy, editor types) to go over your mission statement many times until every word sizzles. Your mission statement should be error-free, eloquent, and precise. It should be dynamic and inspirational. In short, it should be as close to 'perfect' as you can get it.

- **Spread the word**

  Once your mission statement is complete, start sharing it by posting it everywhere you can. It should be prominently displayed on the company's web site, as well as in brochures and other marketing collateral. You can even consider adding it to the bottom of company e-mails or sending it out as a press release. Be creative in getting the word out.

- **Revise as needed**

  Your mission statement, as wonderful as it might sound now, should not be set in stone. As your business grows and changes, so too might your company's mission. Revisit your mission statement on a regular basis to evaluate whether it should be revised or updated. If you've hit the nail on the head the first time around, you probably should not need to alter it significantly as time goes by.

Using the suggested guidelines above develop a mission statement for your organisation. Bear in mind the three areas of organisational beliefs, organisational values and business definitions.

3 Core competences of the organisation

3.1 Identifying core competences

The most powerful way to prevail in global competition and particularly in challenging and uncertain times is to be clear about the roots of your competitive advantage. The organisations to emulate are those that are adept at reinventing themselves, inventing new markets, quickly entering emerging markets and dramatically shifting purchasing patterns in established markets.

Anheuser-Busch

The maker of Budweiser, the king of beers. What is Anheuser-Busch's core competency? Is it the taste of its beer, the power of the advertising, the ability to introduce new products, or its sheer size? Ask anyone in the beer industry who competes against the giant and they will tell you very quickly it's their distribution system. The beer producer has a system of experienced distributors, who solely distribute the Budweiser product to retailers and taverns. The outcome is that Budweiser gets total attention and support and doesn't have to 'fight' to get the attention of their distributors. As a result, Budweiser can introduce new products, get shelf space with retailers, and execute marketing programs far more efficiently and effectively. Its competition isn't so lucky. They often split distributors, with the result that powerful distributors sell multiple brands. The King of Beers is the king, not because of its product, but because of the unique and powerful distribution system it has created. This might seem easy to achieve but is deceptively difficult. Often it requires radical change in the management of organisations. If an organisation is in decline it is important that the senior management team assume responsibility.

In the short term, an organisation's competitiveness derives from the price/performance attributes of the current range of products or services. In the long run, competitiveness derives from an ability to build, at lower total cost and more speedily than competitors, the core competences that spawn unanticipated products.

The real sources of competitive advantage are found in management's ability to consolidate organisation-wide technologies, production skills, knowledge and learnings into competences that enable the organisation to adapt quickly to emerging opportunities and fulfil its mission.

Definition

Core competences: 'the collective learning in the organisation, especially how to co-ordinate diverse production skills and integrate multiple streams of technologies and organising work and the delivery of value'.

Key core competences are communication, involvement and a deep commitment to working across organisational boundaries. Large organisations may involve many levels of staff across a wide variety of

functions who together have the means and legitimacy to blend their own expertise with those of others in new and interesting ways.

Unlike physical assets which deteriorate over time, competences are enhanced as they are applied and shared. Competences need to be nurtured and protected.

Prahalad and Hamel (1990) describe competences as 'the glue that binds existing businesses' and 'the engine of new business development.' Patterns of diversification are guided by them, not just by the attractiveness of markets.

3.2 Identifying core competences – and losing them

So how do you identify a core competence?

- A core competence provides potential access to a wide variety of markets.

- A core competence should make a significant contribution to the perceived customer benefits of the end product.

- A core competence should be difficult for competitors to imitate – and it will be difficult if it is a complex harmonisation of individual technologies and production skills.

Few organisations are likely to build world leadership in more than five or six fundamental competences. If a company complies a list of 20 to 30 capabilities then it has probably not got a grasp of its core competences. However, creating a long list is probably a good start and will help to identify and bundle similar capabilities together.

Discussion

Study buddy

Using the above guidelines create a long list of the core competences of your own organisation. Based upon this, shortlist this to five or six core competences. Compare and contrast your list with your Study buddy.

Those organisations that ignore identifying and building on their core competences do so at their peril. Organisations judging their own and their competitors' competitiveness primarily in terms of price/performance are encouraging the erosion of core competences – or are making too little effort to enhance them.

While it is possible for such organisations to have a competitive product when the fundamental technologies, environmental conditions or business models change through external forces, that organisation could also be vulnerable.

While activities such as outsourcing can provide a shortcut to building a more competitive product this is often done at the risk of building the people-oriented skills required inside the organisation to sustain product leadership.

If an organisation chooses to pursue alliances, partnerships and joint ventures in order to achieve and maintain competitive advantage, this can only be done once it has decided where to build its core competences.

Another reason for losing competences is to forego opportunities to establish competences that are evolving in existing businesses. In divesting of businesses too early, organisations run the risk of closing the door on future opportunities.

3.3 From core competences to core products

The link between core competences and the end product or service is termed the core products. These are the physical embodiment of one or more core competences.

Diversified organisations have a portfolio of products and a portfolio of businesses. Such organisations also have a portfolio of competences as well. Senior management must possess the skills, capabilities and vision to build such competences across the organisation and implement the administrative means for assembling resources spread across multiple businesses.

There are a number of risks to establishing sustainable core competences if senior management perceive its organisation as a multiplicity of separate business units:

(a)   **Under investment in developing core competences and core products**

- No single business may feel responsible for maintaining a viable position in core products or justify the investment required to build world leadership in some core competence.

- Managers of such business units will tend to under invest with a holistic view imposed by senior management.

- It may also result in the organisation's mission being diluted – the business unit may actually choose to work outside the mission statement or create its own.

(b)   **Imprisoned resources**

- As a business unit evolves it develops its own set of competences.

- Typically, the people who embody this competence are seen as the sole property of the business in which they live and grow up.

- Managers of such 'competence carriers' are not only unwilling to lend their talented people to others but may actually hide talent to prevent its redeployment in the pursuit of new opportunities.

- When competences become imprisoned the people who carry the competences do not get assigned to the most exciting opportunities, thus they become demotivated and their skills begin to atrophy.

- Senior management are seldom able to look four or five levels down into the organisation, identify the people who embody critical competences, and move them across organisational boundaries.

(c)   **Bounded innovation**

- If core competences are not recognised the business units will only pursue those innovation opportunities close at hand.

- Hybrid opportunities such as laptops, notebooks and 'iPods' will only emerge when managers take off their business unit 'blinkers'.

- Conceiving of the organisation in terms of core competences widens the domain of innovation.

	Business unit	Core competence
Basis for competition	Competitiveness of today's products	Inter-firm competition to build competences
Corporate structure	Portfolio of businesses related in product-market terms	Portfolio of competences, core products and businesses
Status of the business unit	Autonomy is sacrosanct, the business unit 'owns' all resources other than cash	Business unit is a potential reservoir of core competences

	Business unit	Core competence
Resource allocation	Discrete businesses are the unit of analysis; capital is allocated by business	Businesses and competences are the unit of analysis; top management allocates capital and talent
Value added of top management	Optimising corporate returns through capital allocation trade-offs amongst businesses	Enunciating strategic architecture and building competences to secure the future

Two concepts of corporation (De Wit and Meyer, 2005)

3.4 Redeploying to exploit competences

In order for an organisation to maximise the ability for its core competences to fulfil its organisational mission and ensure business units share competences in the common good senior management can implement a number of measures.

With the help of stakeholders within the organisation, in particular business unit managers, identify organisational-wide competences and ask managers to identify the projects and people associated with them. This sends an important message to middle management – core competences are corporate resources and as such may be reallocated by corporate management on products and projects that enable the fulfilment of the corporate mission. This can be further underlined if each year during the strategic planning or budgetary process, business unit managers must justify their hold on the people who carry the organisation's core competences.

The positive contribution of a business unit manager should be made visible across the organisation. Such co-operative managers should be celebrated as team players.

'Competence carriers' should be regularly brought together from across the organisation to trade notes, ideas and share best practice. This will help build a strong sense of community.

From core competences new business opportunities are born and developed. Only if core competences are known, nurtured and developed can an organisation ensure the successful fulfilment and development of its organisational mission and be best prepared for the future.

3.5 Strategy as the exploitation of competences

Strategic opportunities must be related to the firm's resources. A strategic approach involves identifying a firm's competences.

Definition

A **core competence** is the distinctive competence of an organisation is what it does well, uniquely, or better than rivals.

Core competences critically underpin the organisation's competitive advantage, Johnson & Scholes (1998). According to Johnson & Scholes (1998) an organisation must achieve at least a threshold level of competence in everything it does.

The organisation's core competences are those where it outperforms competitors and that are difficult to imitate.

Competitiveness depends on unique resources or core competences. The organisation's level of performance in its core competences may be judged in three ways.

(a) Comparison with past results
(b) Comparison with industry norms
(c) Benchmarking

3.6 Tests for identifying a core competence

A core competence:

- Provides potential access to a wide variety of markets. GPS of France developed a core competence in 'one-hour' processing, enabling it to process films and build reading glasses in one hour.

- Contributes significantly to the value enjoyed by the customer. For example, for GPS, the waiting time restriction was very important.

- Should be hard for a competitor to copy. This will be the case if it is technically complex, involves specialised processes, involves complex interrelationships between different people in the organisation or is hard to define.

In many cases, a company might choose to combine competences.

Bear in mind that relying on a competence is no substitute for a strategy. However, a core competence can form a basis for a strategy.

GMN viewpoint

Competency dependence

It is important to heed the warnings of business history. GMN Global Honorary President Dr Jagdish Sheth (2007) in his book *The Self Destructive Habits of Good Companies* warns of an organisation's over dependence on its existing core competences.

What do you do when your core competency has become obsolete or non-competitive? What do you do when somebody else is doing a better job? What do you do when all you customers are deserting you? Sheth says that if you feel trapped you have become competency dependent.

Sheth (2007) identifies four things that lead to competency dependence:

- **R&D dependence**

  If R&D is what you do at the expense of other basic functions such as marketing or sales, you run the risk of developing products for which there is no market.

- **Design dependence**

  When your competence must accommodate 'fashion trends' such as clothing then such a dependence can become self destructive.

- **Sales dependence**

  The 'direct sellers' of the past such as Avon who were born in more innocent times need to adapt to changing market conditions by investing more in marketing and R&D.

- **Service dependence**

  Organisations such as travel agents that relied on providing service in the past as a means of differentiation have faced severe competition from online providers and are having to refocus how they deliver service.

Sheth (2007) identifies that the warning signs of competency dependence are:

- Your efforts to transform the organisation have been futile
- The thrill is gone
- Stakeholders are jumping ship

Sheth (2007) offers five remedies:

- **Find new applications**

  For example, the baking soda brand 'Arm and Hammer' has found it can be an environmentally effective cleanser and is now used as a toothpaste.

- **Find new markets**

  Shift the emphasis to emerging markets such as India, China and other ASEA countries. US founded direct-sales organisation Amway's biggest market is now China with sales topping $2bn in 2004.

- **Move upstream, move downstream**

  Expand the range of competences by moving up or down the value chain. An example is IBM which has successfully moved in both directions, providing servers, chips and software to other computer makers while at the same time developing its service business.

- **Develop a new competency**

  Grow your business from a new place through undergoing a transformation. Kodak has had to move from film producer to embracing the digital technology era at a huge $3bn investment cost. It has created deals with such organisations as Intel, Dell and Adobe Systems to maintain its role in the photofinishing and camera department.

- **Refocus your resources**

  Redirect your efforts into areas with more growth and profit potential.

As Sheth concludes a clear wide-ranging vision that opens up possibilities is the first step to breaking the habit of competency dependence.

4 Leadership and management

4.1 Difference between leadership and management?

There are differences in the roles of leaders and managers and it is worth noting what these differences are. Both leaders and managers can and do share skills and both need to sometimes lead and sometimes manage.

However, leaders and managers also have distinct roles that complement each other and one without the other will eventually lead to failure.

Definition

There are as many different definitions of leadership as there are people who have tried to define it (Stodgill, 1974), however, in an attempt to review the existing definitions, Northouse (2009) states:

Leadership '*is a process whereby an individual influences a group of individuals to achieve a common goal' p. 2.*

In doing so, leaders outline what needs to be done by creating and identifying new ideas and trends and establishing a vision and clear direction.

Management works out how to achieve the vision and gets things done through organising, planning, motivating, co-ordinating and controlling resources.

Northouse (2009) argues that management was created as a way to reduce chaos in organisations and utilises the primary functions of management as planning, organising, staffing and controlling.

This is not to say leaders and managers do not work together on both vision and its implementation. As with the best of teams, ideas are explored, practicalities investigated and the process of exchanging ideas and thoughts between leaders, managers and staff is essential. This collaboration and each person understanding the essential role they play is what leads to successful organisation performance.

Understanding the difference is important for knowing where the majority of your time should be spent and what skills should be improved and developed and where you leave or delegate responsibility for aspects of organisation success to someone else.

Many of us will have experienced organisations where the leaders are focused on day-to-day tactical activities at the cost of any sort of strategic direction and where managers dabble in visionary ideas neglecting or relinquishing responsibility for the design and management of implementation.

The result can often be chaos, even in very small organisations. The exception in very small organisations is that the leader and manager may be the same person and here the challenge is often either having the right skills or knowing when and what role to play.

Global case study

Some quotes on leadership and management may help you to consider the differences between leadership and management:

"You manage things; you lead people." – Grace Murray Hopper

"As we look ahead into the next century, leaders will be those who empower others." – Bill Gates

"Leadership and learning are indispensable to each other." – John F. Kennedy

"People ask the difference between a leader and a boss. . . . The leader works in the open, and the boss in covert. The leader leads, and the boss drives." – Theodore Roosevelt

"Managers have subordinates – leaders have followers." – Murray Johannsen

"Successful leaders don't start out asking, "What do I want to do?" They ask, "What needs to be done?" Then they ask, "Of those things that would make a difference, which are right for me?" They don't tackle things they aren't good at. They make sure other necessities get done, but not by them. Successful leaders make sure that they succeed! They are not afraid of strength in others." – Peter Drucker

"I suppose leadership at one time meant muscles; but today it means getting along with people." – Indira Gandhi

"Co-operation with others. Perception, experience, tenacity. Know when to lead and when to follow" – Deng Ming-Dao

"Leadership is influence. It is the ability to obtain followers. When the leader lacks confidence, the followers have no commitment. A leader is great not because of his power, but because of his ability to empower others" – John Maxwell

"The leader has to be practical and a realist; yet he must talk the language of the visionary and the idealist." – Eric Hoffer

---

An interview with Peter Drucker on leadership by business magazine Forbes can be accessed here: http://www.forbes.com/2004/11/19/cz_rk_1119drucker.html.

Discussion

Discuss with your study buddy the key differences between leaders and managers and share your thoughts in the discussion forum.

Now give an example from your own experience of good leadership and contrast it with an example of good management. Likewise with poor leadership and management.

What learning points about the characteristics of good leaders and good managers do you draw from this exercise?

How could you set about developing your own leadership and management capabilities to enable you to benefit from this learning?

What will you actually commit to doing differently?

5 Strategic leadership theories

There are still some prevailing myths of leadership it is worth dispelling.

Leaders are born not made	There are no doubts some people have certain characteristics that predispose them to being effective leaders.
	However, we can all learn and do all the time from the examples of leadership we experience early on of what makes good and bad leaders.
	If we have a personal interest in being a good leader we automatically adopt good practices and try to avoid bad leadership.
Leaders must have charisma	If this were true we would be seriously short of leaders. More importantly, it supposes that charisma automatically leads to good leadership which is not true.
	There are charismatic people that are poor at leadership, they may be able to inspire but lack the skills to do anything positive or productive with inspired people.
	This is not to say that charisma, in a good leader, is not an advantage, it certainly is and used appropriately can be very powerful. However, there are good leaders that do not have charisma and are successful.
Leaders can only lead in one particular way	There are some people who are inflexible and cannot adapt to different situations and usually these people do not make good leaders. We also cannot change our personality to be someone different.
	However, any effective leader can be flexible and modify their leadership style to reflect the requirements of the situation.
Leaders need status to lead	This is often used inappropriately as a crutch for employee compliance, status (job title, formal position higher up the hierarchy) is seen as all that is needed to lead people.
	If this were true we would never experience those occasions when some people take control of a difficult situation or crisis regardless of their status and lead people to a successful outcome, often seen during a natural disaster.
	In the workplace there are examples of a leader's assistant or deputy being seen as the force behind the leader and employees regard them as the real leader.

<table>
<tr><td>Leaders have power and authority</td><td>This is often true as it comes with the territory so a job title or formal position higher up the hierarchy bestows a certain amount of power and authority.

However, if a leader is not trusted or respected this power and authority can quickly diminish, even disappear. Real power and authority comes from the trust and respect of others, employees and other stakeholders. This trust and respect is only conferred on leaders who are seen to be effective leaders.

Leaders may all lead in their own very distinct and different ways but effective and successful leaders often share some common characteristics.</td></tr>
</table>

Global case study

Jack Welch has been with the General Electric Company (GE) since 1960. Having taken GE with a market capitalisation of about $12 billion, Jack Welch turned it into one of the largest and most admired companies in the world, with a market value of about $500 billion, when he stepped down as its CEO 20 years later, in 2000. Although Jack Welch is "the celebrated leader of a global manufacturer often noted for its technological prowess, he has utilised a very human process to drive change through GE's vast organisation. Having respect for the individual as a pivotal force in organisational change, Welch created a model of exceptional performance every corporate leader can learn from. As Jack Welch wrote in a letter to shareholders:

"In the old culture, managers got their power from secret knowledge: profit margins, market share, and all that... In the new culture, the role of the leader is to express a vision, get buy-in, and implement it. That calls for open, caring relations with every employee, and face-to-face communication. People who can't convincingly articulate a vision won't be successful. But those who can will become even more open – because success breeds self-confidence."

Welch believed that great business leaders have to possess large doses of energy, and know how to use that energy to energise others.

Welch moves from meeting to meeting, conveying that message – and a host of other ones as well, some of which have become his trademarks:

- Business is simple
- Don't make it overly complicated
- Face reality
- Don't be afraid of change
- Fight bureaucracy
- Use the brains of your workers
- Discover who has the best ideas, and put those ideas into practice

Jack Welch summed up his prescription for winning in three words:

- Speed
- Simplicity
- Self-confidence

Under Welch's leadership, managers had wide latitude in building their GE units in entrepreneurial fashion. Determined to harness the collective power of GE employees, Jack Welch redefined also relationships between boss and subordinates. He wrote:

"The individual is the fountainhead of creativity and innovation, and we are struggling to get all of our people to accept the countercultural truth that often the best way to manage people is just to get out of their way. Only by releasing the energy and fire of our employees can we achieve the decisive, continuous productivity advantages that will give us the freedom to compete and win in any business anywhere on the globe."

http://www.1000ventures.com/business_guide/crosscuttings/cs_leadership_welch.html
Accessed 21 March 2010

5.1 Leadership characteristics

You do not have to be in the workplace for very long before you can identify what characterises good and bad leaders. You will also have observed the impact certain leadership styles have on you and others. It can be positive and inspiring, negative and destructive or somewhere in between. The traits typically found in successful leaders include:

- Self confident and energised
- Tolerant of stress and willing to assume responsibility
- Persistent and determined
- Decisive, assertive and fair
- Adaptable and co-operative
- Humility

These traits are complemented with the most typical skills of successful leaders:

- Fluent and articulate
- Socially skilled
- Organised and co-ordinated
- Conceptually skilled and creative
- Diplomatic, tactful and persuasive

Many people may start with some innate traits and skills but most leaders develop skills throughout their careers and usually because they want to be leaders and successful leaders, and success does not necessarily mean financial.

The trait theory of leadership proved to have shortcomings when it came to defining leadership not least because leaders are not necessarily born but can be made. Focusing on ideal leadership traits exposed that people with these particular traits did not always make good leaders and not having all of these traits early on did not always stop people from becoming successful leaders.

In particular, it was recognised that successful leaders are not just intelligent and technically skilled. Increasingly emotional intelligence is seen as essential to good leadership.

Discussion

Study buddy

Share with your study buddy who you most admire as a leader and why. Share your examples in the discussion forum.

5.2 Emotional intelligence

Definition

Emotional intelligence is 'the ability to monitor one's own and others' feelings, emotions, to discriminate among them and use this information to guide thinking and actions'.

Mayer and Salovey (1990) described four key factors:

- **Identify emotions**: the ability to perceive emotions in oneself, others, and in objects, art and events.

- **Use emotions**: the ability to generate, use and feel emotion to communicate feelings, or employ them in thinking or creating.

- **Understand emotions**: the ability to understand emotional information, how emotions combine and progress, and to reason about such emotional meanings.

- **Manage emotions**: the ability to regulate emotions in oneself and others to promote personal understanding and growth.

It was the 1990s before the concept became popular in leadership writing on the publication of Daniel Goleman's book (1995) *Emotional Intelligence: Why it can Matter More than IQ*. Goleman originally identified numerous factors under two headings but these have evolved into the following:

- **Self-awareness**: emotional self-awareness, accurate self-assessment, self-confidence.

- **Self-management**: emotional self control, transparency, adaptability, achievement orientation, initiative, optimism, conscientiousness.

- **Social awareness**: empathy, organisational awareness, service orientation.

- **Relationship management**: inspire, influence, develop others, change catalyst, conflict management, building bonds, teamwork, collaboration, communication.

Goleman (1995) believes that emotional intelligence competences are not innate skills but rather they are abilities that can be learned.

He also suggests that you cannot teach emotional competences using traditional methods designed for cognitive learning.

Emotional learning involves thinking and acting in ways that are fundamental to a person's identity, their personality. This will inevitably bring some people into conflict with their deep-rooted beliefs and attitudes and the need to change their behaviour. This can be difficult, it requires new ways of thinking and acting.

Higgs and Dulewicz (2002) identified seven elements of emotional intelligence and broke down the elements into three areas:

- **Drivers**: motivation and decisiveness, traits that energise people and drive them towards achieving goals.

- **Constrainers**: conscientiousness, integrity, emotional resilience, factors that control, curb the excesses of the drivers.

- **Enablers**: sensitivity, influence and self-awareness, traits that facilitate performance and help individuals to succeed.

Higgs and Dulewicz (2002) argue that their components of emotional intelligence divide into two categories for learning.

- People can learn through established methods, such as personal development strategies, eg sensitivity, influence and self-awareness.

- More enduring elements of an individual's personality that are more difficult to learn, such as motivation, emotional resilience and conscientiousness, can be learnt through training strategies that exploit each individual's characteristics to the full and through developing 'coping strategies' that minimise the impact of potential limitations.

Appraising emotional intelligence is best undertaken through a combination of methods including self-assessment, peer assessment, manager assessment and subordinate assessment. This should produce a well-rounded perspective and help identify areas for improvement.

5.3 Leadership styles

Much has been written on leadership styles over the last six decades and many schools of thought have emerged.

> ## Definition
>
> **Leadership style** is the outward expression of the beliefs, values and assumptions held by an individual that lead to typical attitudes and behaviour towards others and influence approaches to leading and managing people.

The journey and evolution of leadership styles

Renaissance

Behavioural (1950s–1960s)	Situational (1970s–1980s)	Transformational (1990s–2000s)	Entrepreneurial (centuries old)	Servant / stewardship (centuries old)

Leadership and management styles vary considerably and the following covers some examples of better known and more enduring theory. Many styles have their value as well as their disadvantages. Leadership styles come in and go out of fashion and even come back into fashion. The leadership styles detailed below emerged in different eras but are still practiced and therefore still relevant today.

5.3.1 Behavioural leadership

Behavioural leadership theory focused on what people did rather than what traits they had. Different ideas emerged but essentially focused on four main themes:

(a) **People-oriented** leadership: focuses on people to achieve the objectives
(b) **Task-oriented** leadership: focuses on tasks to achieve the objectives
(c) **Directive** leadership: authority and decisions are retained by the leader
(d) **Participative** leadership: authority and decisions are shared with employees

Tannenbaum and Schmidt (1973) developed their continuum of possible leadership styles illustrating the degree of control exercised by leaders and extent of freedom for subordinates to participate. They believed that the leadership style adopted depended on three forces:

(a) **Forces in the manager**: eg values, attitudes, personality, behaviour

(b) **Forces in subordinates**: eg personality, need for independence, willingness to take responsibility

(c) **Forces in the situation**: eg environmental factors such as internal (organisational, the problem, time) and external (challenges)

The assumptions of behavioural leadership that leaders either focus on people (goal accomplishment) to achieve goals or focus on tasks (goal attainment) suggests, for example an autocratic style of leadership would tend to focus on task and assume people need to be directed and controlled if the task is to be achieved. A more democratic style of leadership would assume that people can be trusted to complete the task and therefore it is more important to focus on people and their needs. Behavioural leadership can and expects to adapt and change depending on circumstances.

Dictatorial	**Autocratic**		**Democratic**				*Laissez faire*
Manager makes decision and enforces it	Manager makes decision and announces it	Manager sells decision	Manager presents ideas and invites questions	Manager presents tentative decisions subject to change	Manager presents problem, gets suggestions, makes decision	Manager defines limits; asks group to make decision	Manager permits subordinates to function within limits defined by superior

High degree of control exercised by leader
Low employee involvement

Low degree of control exercised by leader
High employee involvement

Adapted from Tannenbaum and Schmidt (1973)

Blake and Mouton (1985) developed their Managerial/Leadership Grid, a two dimensional grid using 'concern with people' and 'concern with production' as the axes. They later added another axis 'motivation'.

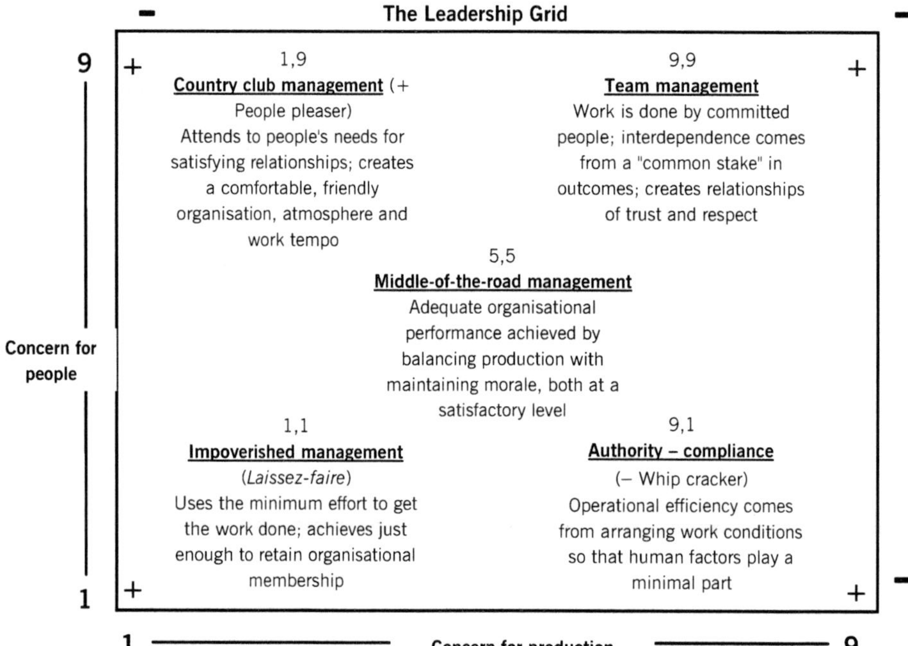

The concern refers to the leader's attitudes, assumptions and style of leadership and the grid reflects how the leader expresses concern for people or concern for production.

The 'term' concern reflects the emphasis that is suggested by respondents to a questionnaire, which identifies five potential leadership styles:

Impoverished management

- Low concern for people and production
- Remote from subordinates
- Little interest in achieving organisational goals or vision

Authority – compliance management

- High concern for production but low concern for people
- Reliance on the application of standard procedures and policies to determine action rather that the possible contribution of staff in the achievement of objectives

Country club management

- High concern for people but low level of concern for production

 Aligning the organisation to the vision

- Concerned with need for harmony and avoidance of conflict, letting subordinates get on with the job

- Seek a comfortable working environment in the belief that production automatically follows

Middle-of-the-road management

- Medium level of concern for people and production

- Keeping everyone happy is a typical approach

- However they tend to underachieve on both counts, achieving neither good levels of production nor highly integrated teams

Team management

- Equally high concern for people and production

- Such leaders create teams in which the needs of individuals and the search for output become integrated

The research of Blake and Mouton (1985) revealed that leaders may change from one style to another or use a combination of different styles for different situations but that leaders typically have one dominant style that usually asserts itself. However, leaders are known to have a "secondary" style that they use if the dominant style is not achieving the desired results.

Styles that leaders can adopt are affected by those they are working with and the environment they are working in. The problem with behavioural approaches to leadership is that they don't look properly at the context or setting in which the style is used.

5.3.2 Situational (contingency) theory of leadership

Situational leadership believes that a leader will adopt a leadership style that is appropriate to the needs of the situation. It is based on the assumption that the person most qualified and experienced to do the job will emerge as leader.

House and Mitchell (1974) linked leadership behaviour with subordinate motivation, performance and satisfaction. They identified four types of behaviour:

(a)   **Directive leadership** – letting people know exactly what is expected and giving specific directions. People expected to follow rules.

(b)   **Supportive** – friendly, approachable, concerned for the needs and welfare of employees.

(c)   **Participative** – consulting and evaluating opinions and suggestions from employees and stakeholders.

(d)   **Achievement-oriented** – setting challenging goals, seeking improvement in performance and confident that subordinates that have ability will perform well.

And two main situational factors:

(a)   **Personal characteristics of subordinates** – determines how they will react to the manager's behaviour and the extent to which they see behaviour as an immediate or a potential source of need satisfaction.

(b)   **Nature of task** – extent to which it is routine and structured or non routine and unstructured.

Hersey and Blanchard (1982) based situational leadership on "*readiness*" (the followers' ability and willingness to accomplish tasks) and the level of people the leader is attempting to influence.

The leadership style reflects the amount of direction and support that the leader provides to their followers. They describe leadership styles as four types of behaviour:

- **Telling**

  If followers display a low level of readiness to be willing and able to achieve the task then the leader should adopt a task-oriented style by telling them what is expected.

- **Selling**

  Most appropriate where followers display moderate levels of readiness towards the task to be achieved.

- **Participating**

  Where medium levels of readiness are found it is possible to favour a relationship situation in terms of style.

- **Delegating**

  Where there are high levels of readiness there is opportunity to delegate much responsibility. The leader becomes more facilitator than manager.

Critical difference	
Transactional / situational	**Transformational**
'Do what is required'	*'Go the extra mile'*
Leader adapts to various situations	Organisation adapts, focus on consistency
Focus in on day-to-day activities	Strategic emphasis, vision
Development of task and people behaviours	Symbolic influence
Situational analysis decides emphasis	Focus is on the organisation
Management by exception (***Active***: seek deviation, correct. ***Passive***: correct if exposed)	Inspire individuals to transcend self-interest, activate higher level needs
Exchange rewards for effective performance	Growth and development of individuals

No one style is considered the best or even necessary for all leaders. Because personality leaders will favour a particular style the challenge for effective leaders is to be flexible and able to adapt to the situation and assume a more appropriate leadership style.

5.3.3 Transformational leadership theory

Transformational leadership is based on transforming organisation performance and such leaders are willing to make changes where necessary to improve and achieve the organisation's goals.

The focus of the transformational leader is to create a vision that appeals to the values of followers and in so doing create a feeling of justice, loyalty and trust.

Transformational leadership can be democratic but the leader expects to shape and change events rather than adapt to changing events which can result in a more autocratic style.

They are focused on strategic issues and believe charisma, intellectual stimulation, inspiration and using motivation will get things done.

Transformational leaders challenge and assume people can be inspired to achieve high standards of performance. Learning and development is therefore a driving force. However, they can be inflexible, single minded, obstinate, stubborn.

As has already been stated following the two leadership styles are actually centuries old but have become increasingly relevant to a broader range of organisations driven by social, environmental and ethical goals and by the urgent need for organisations to nurture creativity and exploit innovation. Attention has therefore refocused on these styles and research and studies resulted in new emerging themes.

Tepper and Piercy (1994) found transformational leaders achieved higher levels of acceptance of objectives than those leaders who focused more on making routine decisions and monitoring performance.

5.3.4 Entrepreneurial leadership

Entrepreneurial leadership is recognised as desirable in both small and large organisations and yet, particularly in large organisations, seems elusive.

Ettinger (1983) suggests that there are two types of entrepreneur:

- **Independent entrepreneurs**

  More concerned with independence than power and therefore more likely to create, develop and seek to retain control over their own organisations.

- **Organisation makers**

  Have a stronger need to exercise power than to search for independence.

  They therefore seek to take control of organisations with growth potential and develop them. The result is they rise to the top of the organisation as they drive it forward, obtaining even greater amounts of power.

Global case study

Results of research by Accenture provide some interesting insights. Many CEOs believe the drive for operational efficiency has stifled entrepreneurial behaviour. Almost seven in ten CEOs believe their staff lack the entrepreneurial instinct but at the same time many are worried about losing control if employees become entrepreneurial and half worry that staff can become too entrepreneurial. It would have been interesting to know what they thought "too entrepreneurial" means.

Entrepreneurial leadership has been around for a very long time. The problem is that in recent decades, particularly as the leadership trends we have discussed emerged and influenced leadership styles, entrepreneurial leadership was sidelined and seen as the domain of start-up and small businesses.

Creativity and innovation feature high on the list of entrepreneurial characteristics. The ability and willingness to challenge current thinking and ways of doing things and come up with new ideas is part of the entrepreneur's makeup.

The entrepreneurial instinct is not seen as a threat or the domain of only the leader. It is seen as desirable in many people in the organisation and actively encouraged. The leader should not see themself as the only person who can be the entrepreneur but be prepared to lead and facilitate entrepreneurship. This, of course, means giving up control and being willing for others to take risks and make mistakes.

Developing strategy and building operational efficiency is essential to organisation success and to realising entrepreneurial output and there is no good reason why efficiency should stifle the entrepreneur. As long as organisational design remains flexible and bureaucracy does not take over entrepreneurship can flourish. It is, however, important to clarify entrepreneurial behaviour.

Entrepreneurs challenge and change things. They do not rigidly follow rules but they also do not break them unless for good reason, for example, constraints on organisation success, they follow tried and trusted principles that work to ensure effectiveness and efficiency.

Entrepreneurs also do not instigate change for no valid reason, there is always a well thought through purpose.

Neal Thornberry's (2006) book *Lead Like an Entrepreneur* explores and describes some of the factors and behaviour of entrepreneurial leadership. He distinguishes between transformational and entrepreneurial leadership as outlined below.

Entrepreneurial leadership – focus and roles

	Internal	External
Activist	Miners (Operational value chain)	Explorers (Market)
Catalyst	Accelerators (Unit)	Integrators (Enterprise)

(Row labels *Activist* and *Catalyst* sit under the axis label "Entrepreneurial role".)

Key characteristics of the entrepreneur include:

- **Internal locus of control**: most entrepreneurs have strong internal locus of control to determine their own fate

- **Tolerance for ambiguity**: chaos, uncertainty and disorder comes with the territory

- **Willingness to hire smarter people**: accept personal limitations and seek to balance

- **Consistent drive to create, build or change**: driven by challenge rather than money

- **Passion**: enthusiastic even obsessive.

- **Optimism**: glass half full

- **Sense of urgency**: impatience, don't waste time, miss the boat

- **Perseverance**: not sidetracked or derailed.

- **Resilience**: accept mistakes, learn, bounce back

- **Sense of humour about oneself**: being prepared to accept flaws

Thornberry (2006) describes two dimensions to entrepreneurial orientation.

- **Activist orientation**
  Adopts a direct driver/owner orientation towards value creation, they push others in new directions, they are serial changers, spotlighting flaws and factors that will hold the organisation back.

- **Catalyst orientation**
  Tend not to be direct drivers of opportunity. They help to set up or induce conditions to allow innovation and entrepreneurial opportunities to be consistently and persistently pursued. They are often cultural value setters, believe in innovation, taking risks, allow mistakes and encourage continual learning. They build a structure and climate where others can create value.

Thornberry goes on to describe commitment to entrepreneurship and attitudes to control and how this influences the way an entrepreneur might behave.

In the diagram blow, the horizontal axis refers to how much organisational commitment the entrepreneur must pursue. The less control they have the more they must negotiate, sell, persuade others to achieve their objectives.

1   **Accelerator** – internally-focused leadership but indirect entrepreneurs, instead focusing on how to get people to be more creative, innovative, stimulate different thinking and behaviour.

2   **Miner** – operationally-focused leadership, creating, designing operations to deliver new value propositions.

3   **Explorer** – market-focused leadership, developing new markets or products or both.

4   **Integrator** – organisation level focus developing an organisation-wide entrepreneurial strategy. They are focused on the external environment and on creating an internal environment for opportunism.

Different types/different requirements

1   **Accelerator** – typically need little permission as they run their own unit and have more control.

2   **Miner** – usually need permission and have less personal control because they embark on internal opportunities that cross organisational lines.

3   **Explorer** – have to assemble resources and need to enlist the help of senior management. They often have to act like start-up entrepreneurs if they want their goals achieved.

4   **Integrator** – embed entrepreneurship as part of their strategy and build the support, through structure and systems necessary to keep the momentum going.

Entrepreneurial leaders often lose interest when they have achieved their goal and unless they are in an environment where they can activate their vision for the next goal they will move on to new challenges elsewhere. For the entrepreneur it is about new beginnings, risk and change.

Entrepreneurs often utilise a servant leadership style to achieve their goals recognising that success depends on committed people.

5.3.5 Action-centred leadership

The basis of this approach was developed by John Adair (1983) who view leadership as a function of three separate but linked ideas:

- **Achieving the task**
  Includes activities such as planning work to be done, allocating resources and duties, checking performance and reviewing progress.

- **Building and maintaining the team**
  Includes activities such as building team spirit and maintaining morale, maintaining discipline and training, establishing team leaders.

- **Developing the individual**
  Includes activities such as dealing with the personal problems of subordinates, reconciling the needs of the group and the individual, and the training of individuals.

Although each of these areas of leadership is distinct, actions or inactions in one area influence events in the other two. Adair uses a three-circle model to illustrate the interrelated relationship between the three aspects of leadership.

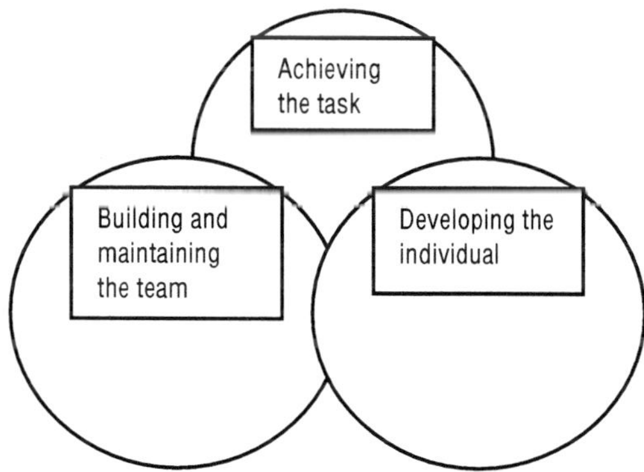

Source: Adair, J (1983) Effective Leadership; A Self Development Manual

5.3.6 Servant/stewardship leadership

Servant leadership goes back thousands of years, with the idea of the leader of a kingdom serving his/her people. It can also be found in the teaching of various religions. The idea of servant leadership in the modern organisation first emerged in 1970 following an essay by Robert Greenleaf on the subject of "The Servant As Leader" where he explained that servant leadership requires a belief that as a leader your role is to serve others first, prioritise the needs of others before focusing on personal needs. Evidence of this servant leadership is found in the personal growth of others.

At the centre of servant leadership is collaboration and the ethical use of power. Stewardship requires sharing power and decision-making.

Greenleaf (1970) identified eleven characteristics of the servant leader:

1   **Calling** to serve in the interests of others and put that interest before self interest

2   **Listening** with commitment to others and to oneself. Hearing and valuing what people say and acting on it when appropriate

3   **Empathy** to understand others in order to better understand their individual concerns and needs

4   **Healing** oneself and others in times of change or crisis, being prepared to understand and allow for reactions of various sorts and work towards solutions

5   **Awareness** of self and others

6   **Persuasion** rather than authority or coercion

7   **Conceptualisation** to cultivate dreams and rise above day-to-day tasks

8   **Foresight** based on learning from the past, comprehend the present and visualise the future

9   **Stewardship** looking after something with the intention of passing it on to others

10   **Growth** of people through formal and informal encouragement of learning and development

11   **Building community** within the organisation and harmonising relationships

The advantages of servant leadership include nurturing a collaborative and supportive culture that encourages personal and organisation growth that facilitates transformation. Its main disadvantage is that such a culture requires time and the commitment of everyone and can lead to a lack of decision-making.

Stewardship is very similar with more of an emphasis on passing something on that is as good or in better condition than when the leader took over, therefore a much longer-term perspective, it protects continuity.

A leader is the guardian of the organisation during their leadership and has a duty to at least maintain, but ideally improve, the strength and performance of the organisation during their stewardship. Stewardship goes beyond fiscal performance and requires attention to broader economic impact, community responsibilities and social inclusion and governance. These issues are currently being addressed by some organisations perusing the goal of corporate responsibility.

Pivotal to stewardship is accountability and stakeholders of many organisations would benefit from leaders being more accountable beyond share price or financial performance.

We might be forgiven for thinking too many leaders have a narrow transient approach to corporate responsibility and are too preoccupied with self-interest. It is interesting that many CEOs argue that it is more difficult to adopt a stewardship style of leadership in a public company.

This argument is weakened with shareholders increasingly demanding a more long-term, responsible approach to business and one of the biggest investment growth areas is in ethical investment. Many leaders would benefit from addressing at least some of the stewardship principles in leading an organisation and increasingly good leaders are doing just that.

Servant and stewardship leadership has been more easily embraced, as a concept at least, by public sector organisations, charities, institutions and religious organisations. However, the re-emergence of

corporate responsibility, as a corporate goal influencing organisation design and leadership behaviour, has seen a renaissance of servant leadership and stewardship thinking.

For many small- to medium-sized organisations, particularly family run businesses, servant stewardship leadership can be automatic with a desire to see their business thrive and continue, whether passed on to the next generation of family or to managers.

All these leadership styles will include a particular way of relating to and motivating people. For example, an autocratic style of leadership assumes people need to be directed and controlled. If the national cultural characteristics support this assumption it may well be true.

However, if people are motivated by taking responsibility and having some autonomy they will be de-motivated by such a style.

Now you have read examples of different leadership style theories identify which styles of leadership are in your organisation. Which leadership style best describes yours? What are the advantages and disadvantages and how you can you improve your leadership style?

6 Approaches to leadership

6.1 Leadership realities

It is one thing to know and understand good leadership practices, concepts and principles but another to implement them. Leaders might know what they should be doing but the organisation and environmental contexts present challenges to implementing improvements and change.

If, for example, you have an autocratic leader who has been entrepreneurial and successful as a result, but is now losing touch with the marketplace and is not interested in whether or not employees are motivated it is very difficult to convince such a leader to change.

If an employee appears to be impossible to motivate it may be because they are in the wrong job and/or organisation and no amount of effort or doing the right thing is going to change that.

Working in these conditions can be challenging in a good way, particularly when you can realise a change and have a positive effect on the organisation. The sense of achievement can be very rewarding.

It can also be demoralising and frustrating when you cannot effect change. But this is the leadership reality. Even at the top of organisations there are constraints and factors out of the control of the leader, influences and factors that have to be negotiated and navigated. Compromise is part of the way we achieve some of what we want and we live with what we cannot achieve.

Being promoted to a leadership role is often the ultimate goal and reward for many executives and they usually work hard to achieve that goal. Having achieved it many find themselves with expected challenges in terms of their tasks and objectives but unexpected challenges in terms of making the transition to leadership.

The Chartered Institute of Personnel and Development (CIPD) research by Development Dimensions International (DDI) on leadership transitions revealed some of the dilemmas leaders face as they move to more responsible roles. All leaders rated the ability to get work done through others as being in their top three most difficult challenges. More than 92% of senior leaders said understanding that the new role would require different ways of thinking would have helped them be successful. One in three said their company provided little or very poor support to them in making the mental shift for the transition. Interestingly, all leaders complained that politics is one of their main challenges as they move to a more senior role rating it as the fourth most difficult challenge.

6.1.1 Leadership paradigms

As we have discussed, leadership styles have come into fashion, evolved, gone out of fashion and become relevant again as the external environment changes.

Leadership style can be dominated by thinking and behaviour of the era, for example when greed becomes acceptable some leaders see no relevance in stewardship and can be driven by narrow goals and self-interest.

New agendas emerge and thinking on what is acceptable leadership changes, for example corporate responsibility is having an effect on what is seen as desirable leadership.

6.2 So what leadership style is right?

There is no one right leadership style, many factors influence what is possible (personal traits and skills), what is right (the situation and circumstances) and what is appropriate (ethical and social goals). As we have discussed leaders need to be able to adapt their leadership style.

Across the hundreds of thousands of different organisations it would be difficult to declare there are dominant leadership styles. All leadership styles described are and will continue to be practiced in varying degrees. An entrepreneurial leader can adopt at one moment a democratic style and at another an autocratic style depending on their personality and the situation.

Modern leadership is showing signs of being more fluid and increasingly reflects a combination of the best of the styles.

There is a thread of servant leadership and stewardship across some leadership styles and as the demand for leaders to become more accountable increases, such styles have become more relevant today.

Taking the best principles from the theory and reflecting on our changing social and business environment can help frame a leadership style that meets a variety of needs and allows for a range of adaptations as appropriate.

Global case study

In an article for African Business, Dr Tunde Ekpe founder of management consultancy Optimentus discussed differences in leadership styles according to cultural issues. Dr Ekpe has carried out research on cultural issues in multinational organisations and she has worked in Russia, Belgium, the Netherlands, Norway, France, Italy, India, Canada, the USA, South Africa, Zimbabwe, Kenya, Botswana, Benin, Ghana, Nigeria, Togo, Senegal and Burkina Faso.

Ekpe stated in the interview:

"You have to work towards becoming a better leader and that in turn means to lead without resorting to fear. There are a lot of people who rule with a very heavy hand. They say 'I am the boss, you do as I say and that's it'. That leaves a lot to be desired with regard to emotional intelligence, the art of interacting effectively with people. These people are perfectly nice people in a social setting and they will tell you about all the work that needs to be done and that being heavy handed is the only way to get it done and stuff like that. I say to them 'yes, but you've got to recognise the other person as a human being as well, and you just never know when you are going to need them again. They might even become your boss tomorrow and then you'll really be in trouble".

In terms of cultural differences she believes that there are some aspects of leadership that are universal, that are common across cultures, that you can evaluate the same way but there are some that are not that easy to evaluate.

She outlines how this works in her consultancy when assessing people according to their leadership style with a view to placing them within a company:

"I am just making generalisations here, but a Japanese candidate for a management role might be a bit more respectful, a bit more reticent about saying 'no' or putting themselves forward, although they never say 'no' really. Then you have differences between Africans. You've got West Africans; they are generally more aggressive and more direct than East Africans who will be more understated."

In terms of her experience and authority to comment on cultural differences she explains:

"I am British and of African descent. Some of the cross-cultural issues that companies face when they are recruiting, I have experienced myself and I can articulate them without embarrassment,".

"It's something of an advantage to have dual nationality and not be restricted by political correctness."

Based on an article for African Business by Williams (2010) which can be downloaded from http://www.optimentus.com/uploads/Profile_tunde_ekpeABno17_0710.pdf

6.3 Leadership strategies

Leaders set the goal for the organisation, the strategic direction and desired competitive position. Strategy is influenced by the leader's personality and their personal goals.

As we have seen an entrepreneurial leadership style is likely to result in being first to market with new products and services, first to exploit new market opportunities. Such leasers are likely to design innovative business models to exploit those opportunities. It will be difficult to keep up with them and they might be exciting to work with but they also may be demanding and difficult.

With entrepreneurs change is continuous and a way of life. An organisation leader with a follower mentality is unlikely to build an organisation that is a market leader (assuming the organisation has the necessary competence).

They may be particularly good at utilising the resources they have very efficiently and effectively and be strong at maintaining follower strategies and a follower position. However, if the organisation has the ability to be a market leader in such circumstances this leader's follower mentality will hold the organisation back.

A leader who believes in stewardship is likely to take a more considered approach to all they do. Their strategy is influenced by the desire to build a strong company for future generations rather than just share price factors and the culture of this organisation will be built on ethical values. Strategy will not compromise short-term goals for longer-term objectives.

The down side is in dynamic markets or periods of unprecedented change where a lack of speed to respond to events can be damaging.

All leadership styles influence the culture of organisations and the culture of organisations influence leadership style.

Activity 5

Evaluate leadership styles in your organisation. Is there a common style, are there different styles and what is the impact on motivation, positive and negative?

6.4 Alternatives to leadership

Kerr and Jermier (1978) suggest three areas where substitution for leadership is possible:

Subordinate characteristics

- Employees are professionally qualified, highly experienced and able to undertake activities expected of them.

- Do not need leadership in the conventional sense of the term.

- Little scope for leadership where the subordinate is indifferent to the rewards the leader can offer.

Task characteristics

- Where work is highly routine and contains immediate feedback on performance and achievement there is little scope for leadership.

- Leadership becomes restricted to ensuring that work is provided to the employee and that it is taken away when it is complicated.

Organisational characteristics

- An organisation that is highly routinised with little flexibility will have limited need for leader activity as it will almost run itself.

1 Understand the importance of an organisational vision

- Leaders and managers must constantly make choices and seek solutions based on an understanding of what their organisation is intended to achieve.

- Where we have a clear understanding of the organisation's vision and mission this can provide strong guidance during processes of strategic thinking, strategy formation and strategic change.

- The successful direction of the organisation is as a result of the effective interaction between management and the board of directors.

- The board of directors must be coherent in its view and implementation of the organisational mission.

- Organisational purpose can be defined as the reason why an organisation exists.

- De Wit and Meyer (2005) identify three other components of a corporate mission as organisational beliefs, organisational values and business definition.

- When a consistent and compelling mission is formed this can ignite the organisation with a sense of purpose.

- There are a number of clear benefits of a clear, coherent and well communicated mission.

2 Compare and contrast perspectives on organisational purpose

- Organisations require a certain amount of economic profitability to survive and they need to exhibit a certain amount of social responsibility if they are to retain the trust and support of key stakeholders.

- A stakeholder grid is an analytical device depicting an organisation's stakeholders.

- Stakeholders need to be involved in decision-making. The downside of involvement is risk.

3 Evaluate the role of core competences in an organisation's mission

- The most powerful way to prevail in global competition and particularly in challenging and uncertain times is to be clear about the roots of your competitive advantage.

- The real sources of competitive advantage are found in management's ability to consolidate competences that enable the organisation to adapt quickly to emerging opportunities and fulfil the organisation's mission.

- Competences are enhanced as they are applied and shared.

- Few organisations are likely to build world leadership in more than five or six fundamental competences.

- The link between core competences and the end product or service is termed the core products. These are the physical embodiment of one or more core competences.

- There are a number of risks to establishing sustainable core competences if senior management perceive its organisation as a multiplicity of separate business units.

- With the help of stakeholders within the organisation managers can identify organisational-wide competences.

- From core competences new business opportunities are born and developed.

- The organisation's core competences are those where it outperforms competitors and that are difficult to imitate.

- Competitiveness depends on unique resources or core competences. They should be hard for a competitor to copy.

- Sheth warns of an organisation's over dependence on its existing core competences.

4 Evaluate the differences between leadership and management

- There are differences in the roles of leaders and managers and it is worth noting what these differences are.

- However, leaders and managers also have distinct roles that complement each other and one without the other will eventually lead to failure.

- Leaders and managers work together on both vision and its implementation. This collaboration and each person understanding the essential role they play is what leads to successful organisation performance.

5 Discuss characteristics of leadership and different strategic leadership theories

- There are some prevailing myths of leadership it is worth dispelling.

- These traits are complemented with the most typical skills of successful leaders.

- The trait theory of leadership proved to have shortcomings when it came to defining leadership not least because leaders are not necessarily born but can be made.

- Increasingly emotional intelligence is seen as essential to good leadership.

- Leadership and management styles vary considerably and the following covers some examples of better known and more enduring theory.

- Leadership styles come in and go out of fashion. Leadership styles emerge in different eras but are still relevant today as they are still practiced.

6 Critically suggest how organisational realities impact on leadership

- It is one thing to know and understand good marketing leadership practices, concepts and principles but another to implement them.

- Leaders might know what they should be doing but the organisation and environmental contexts present challenges to implementing improvements and change.

- There is no one right leadership style. Leadership styles need to be able to adapt.

- Modern leadership is showing signs of being more fluid and increasingly reflects a combination of the best of the styles.

- Taking the best principles from the theory and reflecting on our changing social and business environment can help frame a leadership style.

- All leadership styles influence the culture of organisations and the culture of organisations influence leadership style.

7 Consider appropriate alternatives to leadership

- There are areas where substitution for leadership is possible.

- Substitution for leadership is possible according to Kerr and Jermier (1978) depending on the characteristics of the subordinate, task and organisation.

Activity 1

You will find that intuitively managers will fall into two perspectives; the shareholder value perspective and the stakeholder value perspective. This will be based in part on people's own attitudes, values and behaviours, past experiences and future aspirations.

Do their answers surprise you? An organisation must strive to recruit people that will be positively inclined to fulfil the organisation's mission. If there are major differences in the responses you get question why this is happening? Perhaps senior management has not worked hard enough at communicating its mission? Perhaps people have been misrecruited? How can you ensure that the organisation creates a mission that is widely understood and accepted as possible.

Activity 2

This will depend on your organisation. A Stakeholder Audit Process is a systematic method for identifying stakeholders and assessing the effectiveness of current organisational strategies. By itself each process analyses the stakeholder environment from the standpoint of the organisational mission and seeks to formulate strategies for meeting stakeholder needs and concerns. An analytical device such as a 'stakeholder grid' helps you formulate strategies for coping with them. It has two dimensions. The first dimension is one of 'interest' and 'stake'. The second dimension is its 'power', which ranges from formal power of shareholders, to the economic power of customers to the political power of special interest groups.

Activity 3

De Wit and Meyer (2005) identify three components of a corporate mission as:

Organisational beliefs: To work in unison a common understanding is needed amongst all organisational stakeholders. The stronger the set of beliefs subscribed to by the stakeholders the easier communication and decision-making will become and the more confident, cohesive and driven the group will be. If people do not share the same beliefs joint decision-making will be protracted and conflictual.

Organisational values: Each person in an organisation can have their own set of values. Yet, when an organisation's stakeholders share a common set of values this can have a strong impact on the strategic direction. Widely held values contribute to a clear sense of organisational identity which will attract some individuals and dispel others. To be of most use and influence such values must be embodied in the organisational culture.

Business definition: Most firms have a clear identity from being active in a particular sector or industry. For such organisations, having a delimiting definition of the business they want to be in strongly focuses the direction in which they develop. Their business definition functions as a strong guiding principle and helps to distinguish opportunities from diversions. Here while a clear business definition can focus the organisation's efforts it can also lead to short-sightedness and missing new business developments and opportunities.

Activity 4

An appreciation of different leadership theories may provide you with some insights into your own leadership style. You could also discus this with your peers, your subordinates and your boss. This will enable you to see yourself how others see you. The more you can be open to their comments, some of which may be critical, the more you will improve your own leadership style

Activity 5

Your organisation context will influence your critique of leadership theories but the sort of issues that might emerge are that some styles described are irrelevant, for example if you are in a very creative environment autocratic leadership is unlikely to exist. If you work in a very hierarchical and bureaucratic organisation that operates in a simple and stable environment entrepreneurship may be impossible or seem irrelevant. If you work in a highly competitive, fast moving environment you may think servant leadership or stewardship is appropriate.

Adair, J., (1983). *Effective Leadership: A self development manual*. London: Gower.

Blake, R. R. and Mouton, J. S., (1985). *The Managerial Grid III*. Doha: Gulf Publishing Company.

Campbell and Yeung, (1991). 'Creating a sense of mission'. *Long Range Planning*, Vol. 24, No. 4, August, pp. 10–20.

Collins, J. and Portas, J., (1994). *Built to Last*. London: Harper Business.

De Wit, B. and Meyer, R., (2005). *Strategy Synthesis: Resolving Strategy Paradoxes to Create Competitive Advantage*, 2nd edition. London: Cengage.

Ettinger, J. C., (1983). 'Some Belgian evidence on entrepreneurial personality'. *European Small Business Journal*, January, Vol. 31, pp. 48-56.

Freeman, R. E. & Reed, D. L., (1983). 'Stockholders and stakeholders: A new perspective on corporate governance'. *California Management Review*, 25(3): 93–94.

Goleman, D., (1996). *Emotional Intelligence: Why It Can Matter More Than IQ*. London: Bloomsbury.

Greenleaf, R. K., (1970). *Servant Leadership: a journey into the Nature of Legitimate Power and Greatness*. London: Spears.

Hersey, P. and Blanchard, K. H., (1993). *Management of Organisational Behaviour; Utilising Human Sources*. London: Prentice Hall.

Higgs, M. J. and Dulewicz, V., (2002). *Making Sense of Emotional Intelligence,* 2nd edition. Windsor: NFER nelson.

House, R. J. and Mitchell, T. R., (1974). 'Pack-goal theory of leadership'. *Contemporary Business*, Vol. 3, Fall, pp. 81-98.

Johnson, J. and Scholes, K., (1998). *Exploring Corporate Strategy: Text and Cases*. London: FT Prentice Hall.

Kerr, S. and Jermier, J.M., (1978). 'Substitutes for leadership: Their meaning and measurement'. *Organisational Behaviour and Human Performance,* Vol. 22, (3), pp. 375–403.

Mayer, J. and Salovey, P., (1990). Emotional Intelligence: Imagination, cognition and personality', *Journal of Personality Assessment*, Vol. 54, pp. 772–781.

Northouse, P.G., (2009). *Leadership: Theory and Practice*. Los Angeles: Sage.

Prahalad, C. and Hamel, G., (1990). 'The core competence of the corporation', *Harvard Business Review*, 68, (3), pp. 79–91.

Sheth, J., (2007). *The Self Destructive Habits of Good Companies.* Wharton School Publishing.

Stodgill, R.M., (1974). *Handbook of Leadership*. New York: Free Press.

Tannenbaum, R. and Schmidt, W.H., (1973). 'How to choose a leadership pattern. *Harvard Business Review*, May–June, pp. 162–178.

Tepper, B. T. and Piercy, P. M., (1994). 'Structural validity of the multifactor leadership questionnaire'. *Educational and Psychological Measurement*, Vol. 54, pp. 734-744.

Thornberry, N., (2006). *Lead Like an Entrepreneur*. Oxford: McGraw Hill.

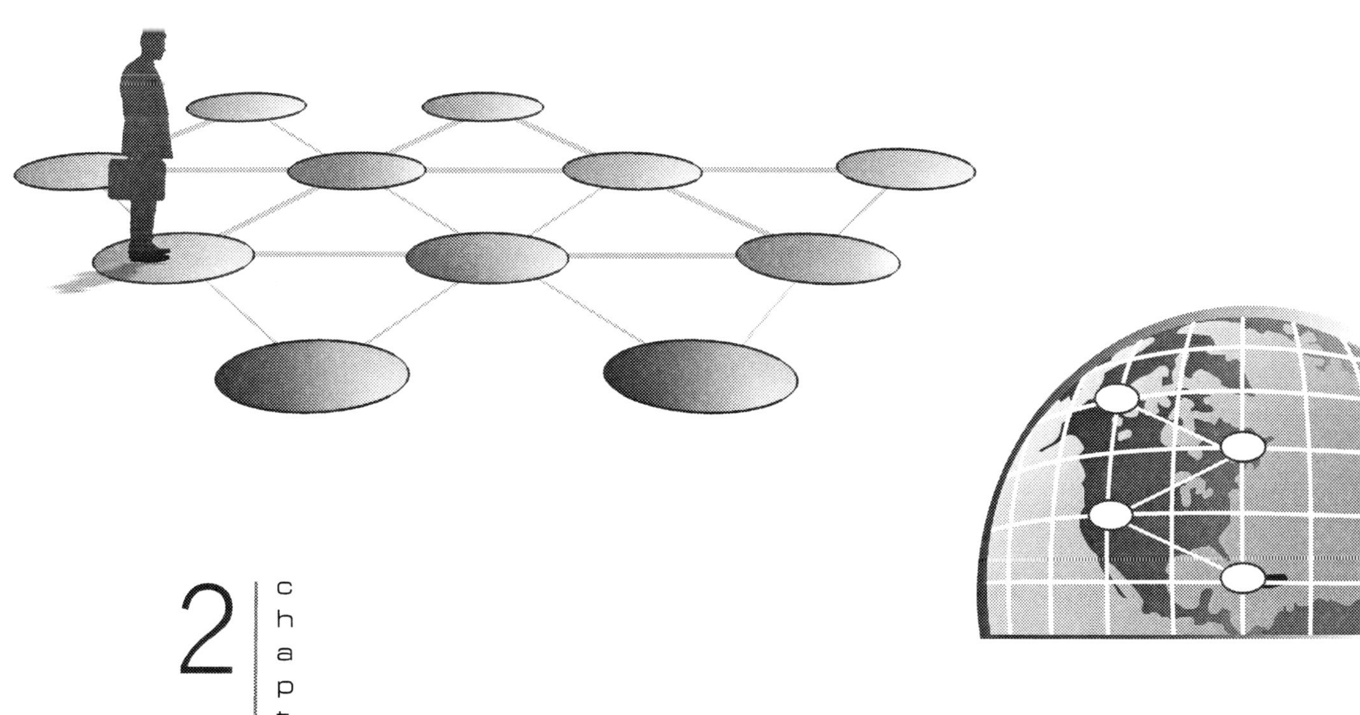

2 | chapter

Motivating groups and teams

This chapter explores a range of topics in the formation of groups and teams and how people are motivated. We start by exploring why team working is so popular while also recognising some of its drawbacks. We look at some key issues in team structure and what kinds of team structures are best able to deliver different outcomes that are focused on the overall vision of the organisation. We also explore how the leader can 'put together' a team with the right mix and balance of competences, and the right contribution structure, to provide a foundation for effective team performance and organisational capability.

In this chapter you will also consider what makes an effective team, and how you might diagnose when a team is ineffective, looking at a number of key processes in team working: the development of teams over time; the emergence of team cohesion or solidarity; group decision-making and group communication. The area of groups and teams in human resource management is a huge area of study: here we are providing you with best-practice frameworks, which you can use as a foundation for your own more detailed reading, analysis and application, both in the workplace and for your assignment.

Contents

Chapter learning outcomes

In this chapter you will cover the following:

- Compare and contrast groups and teams

- Evaluate team structures

- Develop an effective team

- Identify the key elements of motivation

- Evaluate both content and process theories of motivation

- Assess the manager's role in motivation

- Consider the effects of authority, power and influence

1 Groups and teams – are they different?

1.1 Groups

Definition

Handy (1993) defined a **group** as *'any collection of people who perceive themselves to be a group'*.

People are often drawn together into groups by a preference for smaller units where closer relationships can develop; the need to belong and to make a contribution that will be noticed and appreciated; shared space, specialisms, similar objectives or interests; the attractiveness of a particular group activity or resources or access to power greater than individuals could wield on their own (one of the reasons for joining a trade union or staff association, say).

There are two types of groups:

- **Informal groups** will invariably be present in any organisation: examples include workplace cliques and networks of people who get together to exchange information, groups of friends who socialise outside work and so on. Such groups have a constantly fluctuating membership and structure.

- **Formal groups** are intentionally and rationally designed to achieve objectives assigned to them by the organisation, for which they are responsible. They are characterised by: membership and leadership appointed and approved by the organisation; compliance of the members with the group's goals and requirements and structured relationships of authority, responsibility, task allocation and communication.

People contribute different skills and attributes to the organisation as group members than they do as individuals, because:

- Human behaviour is different in groups than in solo or interpersonal situations: group dynamics have an effect on performance.

- Groups offer synergy: such as '2 + 2 = 5' whereby the pooling and stimulation of ideas and energies in a group can allow greater contribution than individuals working on their own.

However, group dynamics and synergy may also be negative: distracting the individual, stifling individual responsibility and flair and so on. Individuals may contribute more and better, in some situations.

1.2 Teams

> **Definition**
>
> A **team** is 'a *small number of people with complementary skills who are committed to a common purpose, performance goals and approach for which they hold themselves mutually accountable*.'
>
> (Katzenbach & Smith, 1993)

All teams are groups – but not all groups are, strictly speaking, teams.

1.2.1 Benefits of using teams

The basic work units of organisations have traditionally been specialised functional departments. In more recent times, organisations have adopted what Peters and Waterman (1982) called 'chunking': the breaking up of the organisation structure into small, flexible units, or teams. From the organisation's standpoint, teams have a number of advantages.

- Teams facilitate the performance of tasks which require the **collective skills**, experience or knowledge of more than one person or discipline.

- Teams facilitate the **co-ordination** of the work of different individuals or groups, by bringing them together across organisational boundaries (eg departments and functions) with shared goals and structured communication.

- Teams facilitate **interactive communication** and interpersonal relationships, and are thus particularly well-adapted for:

    - **Testing and ratifying decisions**, because they offer multi-source feedback and may make the decision more acceptable, by taking account of a cross-section of stakeholder views. Teams have been shown to produce better evaluated (though fewer) decisions than individuals working separately.

    - **Consultation and negotiation**, because they allow an interactive exchange of views and influence.

    - **Generating ideas**, because of their potential for 'bouncing' ideas off each other and getting multiple inputs.

    - **Collecting and disseminating information**, because of the multiple networks in which the various members are involved.

- Teams can motivate their members, since they offer satisfying relationships, mutual encouragement and support, and the opportunity to share work loads and responsibilities. Peters and Waterman (*ibid*) argue that teams enable people to make noticeable individual contributions (which bolsters their **self-esteem**) and at the same time to share responsibility and be part of something bigger than themselves (which bolsters their sense of security).

1.2.2 Drawbacks to using teams

Teams and team working are very much in fashion, but it is important to recognise their potential drawbacks, limitations and challenges.

Team working is not suitable for all jobs: it should be introduced because it leads to better performance, not because people feel better or more secure.

Team processes (especially excessive meetings and seeking consensus) can delay decision-making: groups make fewer decisions than individuals.

Social relationships might be maintained at the expense of other aspects of performance or inter-group conflicts may get in the way of effective collaboration.

Group norms may restrict or inhibit individual contribution. Mayo's Hawthorne studies, for example, showed that individuals restrict their output in order not to 'show up' the team by producing more or better work than the team average (Huczynski and Buchannon, 2001).

Due to a process called social facilitation, performance of simple tasks at which people are relatively confident improves in the presence of other people – but performance of new, complex tasks is hindered by the presence of an audience.

Groups have been found to take riskier decisions than individuals on their own. This may occur due to a number of dynamics.

- **Conformity**: group pressures persuade individuals to agree to decisions, suppressing their questions and 'better judgement'.

- **'Group think'** (Janis, 1972): team consensus and cohesion may prevent consideration of alternatives or constructive criticism. The team becomes blinkered to contradictory information, and feels invincible, leading to ill-considered decisions.

- The **'risky-shift phenomenon'**: if a group is broadly like-minded, discussion will tend to strengthen the prevailing attitude, leading to polarised views (group polarisation). Together with the diffusion of individual responsibility, this can lead to risky decision-making.

- The **Abilene paradox**: group members go along with what they assume is the group's intention – leading to a situation where the group does what none of its individual members really wants to do.

Global case study

Groupthink can be avoided. After the Bay of Pigs invasion fiasco, which was put down largely to groupthink, John F Kennedy sought to avoid groupthink during the Cuban Missile Crisis. During meetings, he invited outside experts to share their viewpoints, and allowed group members to question them carefully. He also encouraged group members to discuss possible solutions with trusted members within their separate departments, and he even divided the group up into various sub-groups, to partially break the group cohesion. JFK was deliberately absent from the meetings, so as to avoid pressing his own opinion. Ultimately, the Cuban missile crisis was resolved peacefully, thanks in part to these measures.

http://en.wikipedia.org/wiki/Groupthink
Accessed 1 April 2010

Activity 1

Consider the use of teams in your organisation? Thinking about specific examples when did a team-based approach prove successful and why? How can your organisation use team working more effectively?

2 Team structure

A team may be called together temporarily, to achieve specific task objectives (eg a project team or task force), or may be more or less permanent, with responsibilities for a particular brand, market segment or stage of the marketing process (eg a brand or process team).

There are two basic approaches to the organisation of team work which requires a variety of skills and competences: multidisciplinary working and multi-skilled working.

 Aligning the organisation to the vision

2.1 Multi-disciplinary and cross-functional teams

Multi-disciplinary and cross-functional teams bring together individuals from different functional specialisms, within a matrix structure or on temporary (project) basis, so that their competences can be pooled or exchanged.

Such teams are an important tool in situations where technical expertise in a range of different disciplines is required. This used to be regarded as primarily the case in large technical projects, but there has also been an increasing focus on horizontal business processes, and the horizontal (cross-functional) structures required to streamline them.

2.1.1 Multi-disciplinary teams

- Increase team members' awareness of the big picture of their tasks and decisions, by highlighting the dovetailing and interdependency of functional objectives.

- Help to generate solutions to problems, and suggestions for performance or process improvements, by integrating more pieces of the puzzle.

- Aid co-ordination across functional boundaries, by increasing the flow of communication, information relationships and co-operation.

- Support stakeholder management and buy-in, by representing the interests and viewpoints of multiple stakeholders in a project or process.

A drawback of multi-functional teams, however, is that their members have dual reporting lines and responsibilities. The team as a whole may represent a wide range of different interests, backgrounds, work cultures, specialist skills, terminology and so on. This creates a particular challenge for the team leader: to build a sense of team identity, role clarity, co-operative working and communication flow

2.1.2 Multi-skilled teams

> **Definition**
>
> **Multi-skilled teams** bring together, in a permanent team structure, a number of functionally versatile individuals, each of whom can perform any of the group's tasks: work can thus be allocated flexibly, according to who is best placed to do a given job when required.

The advantages of multi-skilling:

- Performing a whole, meaningful job is more satisfying to people than performing only one or two of its component operations (as in the kind of job design favoured by scientific management).

- Allowing team members to see the big picture enables and encourages them to contribute information and ideas

- Empowering team members to take initiative enhances the unit's responsiveness to customer demands and environmental changes (particularly in customer-facing units such as sales and customer service).

- A focus on overall task objectives rather than individual job descriptions reduces the need for tight managerial control and supervision.

- Human resources can be allocated more flexibly and efficiently, without potentially disruptive 'demarcation' disputes (about 'whose job' a task is or should be).

Multi-skilled teams are particularly appropriate when a high degree of flexibility is required. They are a cornerstone of team empowerment, since they cut across the barriers of job descriptions to enable teams to respond flexibly to changing demands.

However, there is a limit to how versatile most people can be, so multi-skilling is perhaps best suited to relatively limited areas of activity.

2.1.3 Virtual teams

Virtual teams bring together individuals working in remote locations, reproducing the social, collaborative and information sharing aspects of team working mainly – or wholly – using Information and Communication Technology links.

A virtual team may comprise members in different offices, branches or worksites; on the road (eg sales force teams) or in different countries. Teleconferencing, video-conferencing, locally networked computers, the Internet, mobile telecommunications and other tools can be used to connect remote and mobile team members for data and task sharing, meetings, joint decision-making, performance monitoring and so on.

Localised virtual teams have been used for some time in the form of teleworking: working from home, or from a satellite office close to home, with the aid of PCs, laptops, PDCs, modems and other hardware and software. The main benefits cited for such work include savings on office overheads and the elimination of the costs and stresses of commuting for employees.

More recently, however, the globalisation of business, the need for fast responses to market demands and the increasing sophistication of available technologies has brought about an explosion in global virtual team working. More and more organisations are attempting to conduct business 24–7, with people on different continents and in different time zones.

Global case study

In their book *Uniting the Virtual Workforce: Transforming Leadership and Innovation in the Globally Integrated Enterprise*, Karen Sobel Lojeski and Richard R Reilly cite an insurance company that lost $3 million on one project alone because of issues surrounding virtual work. Some companies, including Hewlett-Packard have even asked that information technology workers begin working in corporate offices again. Yet both argue that recalling workers to the office won't necessarily solve the problem. That's because physical distance is only partly to blame. Virtual distance can also entail operational distance, where there's no shared context among workers in different departments.

Or it can involve what the authors call "affinity distance," which are marked differences in workers' value. Lojeski and Reilly stress that face-to-face meetings are most crucial when first getting a project off the ground, when there are major hitches that need to be discussed openly, and when presenting results to clients. Other suggestions include issuing an e-mail etiquette guide for the team so that everyone, regardless of age or culture, can better manage expectations.

One case study in their book describes how Karan Sorensen, chief information officer for Johnson & Johnson's pharmaceutical research & development, took such steps to manage virtual distance during a global infrastructure project. Sorensen first brought the group together for a face-to-face meeting so that team members could get to know one another. Early on, they set up rules of engagement, such as how each individual liked to communicate best, to address cultural differences. On conference calls, the staff kept photos of everyone by the phone, and Sorensen made sure to alternate call times so that certain people weren't always stuck dialling in at midnight. By getting her team to collaborate better, Sorensen completed the project under budget and well ahead of deadline, saving J&J more than $200 million over three years.

http://www.businessweek.com/technology/content/may2008/tc20080515_505734.htm

Accessed 1 April 2010

2.1.4 Self-managed teams

Self-managed teams are the most highly-developed form of team working. They are permanent structures in which team members collaboratively decide all the major issues affecting their work: work processes and schedules, task allocation, the selection and development of team members, the distribution of rewards and the management of group processes such as problem-solving, conflict management, internal discipline and so on.

The team leader is a member of the team, acting in the role of coach and facilitator: leadership roles may be shared or rotated as appropriate.

Self-managed teams:

- **Contract with management** to assume various degrees of managerial responsibility (which may increase as the team develops) for planning, organising, directing and monitoring. Teams often report to 'absentee' managers with broad responsibilities for several functions, whose role is to act as integrators/facilitators.

- **Perform day-to-day planning and control functions**: scheduling and co-ordinating the daily and occasional tasks of the team and individuals; setting performance goals and standards; formulating and adopting budgets; collecting performance data and reviewing results.

- **Perform internal people management functions**: screening and interviewing candidates to join the team and contributing to selection/hiring decisions; providing orientation for new members; coaching and providing feedback on member performance; designing and conducting cross-training on all tasks.

- **Recruit or cross-train members for multi-skilling**: members rotate or are deployed flexibly from task to task, as required.

- **Use weekly team meetings to identify, analyse and solve task and relationship problems within the team**: reviewing team working and progress; getting team members to research and present team issues and so on.

Self-managed team working is said to have significant advantages in managerial cost savings, and in gains in quality and productivity, through harnessing the commitment and contribution of those who perform the work. However, it obviously presents an additional layer of challenges in developing teams capable of effective self-management.

Global case study

Fisher & Paykel is a New Zealand manufacturer of household whitegoods and healthcare products. Following a period of increased competition from foreign imports in New Zealand, the organisation identified a need to improve team performance. The intention was to excel in creativity amongst all workers and maintain innovation.

'Everyday Workplace Teams' were developed with the vision being that team membership would become an integral part of an employee's everyday work. During the period of change required to integrate teams into the workplace, team leaders were provided with an interactive workbook to help them build team success.

Over time, teams became responsible for organising their own work and took additional responsibilities such as maintenance of their work area. Cost savings were made as a result of fewer reject products and less wastage, better production times and the saving in use of engineers and less downtime from changes to the production process.

Source: based on Mallon and Kearney (2001)

2.2 Team roles and functions

Team membership may be dictated by existing arrangements, organisational appointment or election.

However, where a manager is able to select team members, a mix of attributes, competencis and resources should be secured to match the needs of the task. Team members should be selected for their potential to contribute to getting things done (task performance) and establishing and maintaining good working relationships (group maintenance).

Resources required by a given team may include:

- **Specialist skills**. A team may exist to combine expertise from different functions.

- **Power or influence in the wider organisation** (to help mobilise support and resources).

- **Access to resources**. Team members may contribute information or useful network contacts, or be able to access finance, equipment or staff for the task.

- **Personal attributes** such as commitment, energy or creativity.

2.2.1 Belbin's team role model

R Meredith Belbin (1993) researched business game teams at Henley Management College and drew up a widely-used framework for understanding roles within work teams. Belbin insisted that a distinction needs to be made between:

- **Team (process) role** (a 'tendency to behave, contribute and interrelate with others at work in certain distinctive ways').

- **Functional (task) role** ('the job demands that a person has been engaged to meet by supplying the requisite technical skills and operational knowledge').

Belbin's model of nine roles addresses the mix of team/process roles required for a fully functioning team (Belbin, 1993, p.22).

Belbin's nine team roles

Team role	Characteristics
Plant	<ul><li>Creative, imaginative, unorthodox</li><li>Solves difficult problems, presents new ideas, ignores details</li><li>Too preoccupied to communicate effectively</li></ul>
Resource investigator	<ul><li>Extrovert, enthusiastic, communicative</li><li>Explores opportunities, develops contacts</li><li>Over-optimistic</li><li>Loses interest once initial enthusiasm past</li></ul>
Co-ordinator (or Chairman)	<ul><li>Mature, confident, a good chairperson</li><li>Clarifies goals, promotes decision-making, delegates well</li><li>Can be seen as manipulative</li><li>Delegates personal work</li></ul>
Shaper	<ul><li>Challenging, dynamic, thrives on pressure</li><li>Has the drive and courage to overcome obstacles</li><li>Can provoke others and hurt people's feelings</li></ul>
Monitor–Evaluator	<ul><li>Sober, strategic and discerning</li><li>Sees all options, judges accurately</li><li>Lacks drive and ability to inspire others</li><li>Overly critical</li></ul>

Team worker	• Co-operative, mild, perceptive and diplomatic
	• Listens, builds, averts friction, calms the waters Indecisive in crunch situations
	• Can be easily influenced
Implementer (or Company Worker)	• Disciplined, reliable, conservative and efficient
	• Turns ideas into practical actions
	• Somewhat inflexible
	• Slow to respond to new possibilities
Completer-Finisher	• Painstaking, conscientious, anxious
	• Searches out errors and omissions, ensures delivery on time
	• Inclined to worry unduly
	• Reluctant to delegate
	• Can be a 'nitpicker'
Specialist	• Single-minded, self-starting, dedicated
	• Provides knowledge and skills in rare supply
	• Contributes only on a narrow front, dwells on technicalities
	• Overlooks the 'big picture'

Adapted from Belbin. M. (1993)

Activity 2

Which Belbin team roles do you feel best go with the following phrases and slogans?

(a) The small print is always worth reading.
(b) Let's get down to the task in hand.
(c) In this job you never stop learning.
(d) Without continuous innovation, there is no survival.
(e) Surely we can exploit that?
(f) When the going gets tough, the tough get going.
(g) I was very interested in your point of view.
(h) Has anyone else got anything to add to this?
(i) Decisions should not be based purely on enthusiasm.

Identify the team role for every phrase above.

2.2.2 A balanced team

These team roles are not fixed within any given individual. Team members can occupy more than one role, or switch to 'backup' roles if required: hence, there is no requirement for every team to have nine members.

However, since role preferences are based on personality, it should be recognised that:

- Individuals will be naturally inclined towards some roles more than others.
- Individuals will tend to adopt one or two team roles more or less consistently.
- Individuals are likely to be more successful in some roles than in others.

The nine roles are complementary, and Belbin suggested that an 'ideal' team should represent a mix or balance of all of them.

Effective team working requires a mix and balance of all the roles, which between them support task functions (such as ideas generation, problem solving, implementation and follow-up) and team maintenance functions (support, conflict management, leadership and so on). If managers know employees' team role preferences, they can strategically select, 'cast' and develop team members to fulfil the required roles.

3 Effective and high-performing teams

The task of the leader is to build a 'successful' or 'effective' team. You might immediately think of such a team as one which fulfils its task objectives efficiently and effectively, but there is more to it than that. Adair's (1983) 'action leadership' model and management can be measured by three key criteria:

- **Task performance**: fulfilment of task objectives and marketing/organisational goals.

- **Team functioning**: maintenance of consistent and constructive team working, managing the demands of team dynamics, roles and processes.

- **Team member satisfaction**: fulfilment of individual development and relationship needs.

3.1 Assessing the effectiveness of your team

There are a number of factors, both quantitative and qualitative, that may be assessed to decide whether or how far a team is functioning effectively – over and above whether it is meeting its targets.

What does an effectively functioning team 'look like'?

- The team has a clear mission and objectives which are understood and shared by all members.

- The mix of personalities and skills is diverse and complementary. Members respect each others' differences and appreciate the synergistic effect of diverse individual contributions.

- Each individual is both supported and challenged to contribute to his or her best ability.

- There is a sense of identity and belonging which encourages loyalty and commitment.

- Information and ideas are freely gathered and shared for the use of the team.

- Constructive conflict – new ideas, challenges to the *status quo*, constructive criticism and disagreement – is encouraged, as a way of testing and improving group decisions.

- Each member is encouraged to participate and have a voice in the team, so that (s)he 'buys in' to its decisions and activities.

- There is trust and openness, so that individual and task problems can be safely aired and addressed.

- There are effective mechanisms for maintaining communication, both formal (eg team meetings) and informal (networking and socialising).

- The team accepts and supports the designated leader, but continues to function well and maintain discipline in the leader's absence, and is able to share leadership roles when required.

3.2 High-performance teams

Vaill (1989) identified the characteristics of 'high performance' or extraordinary teams, as follows.

- They perform to a high level against defined external benchmarks or standards, and in comparison with their own previous measured performance (ie they continually improve).

- They perform to a level beyond what is assumed to be their potential best (ie they continually exceed expectations).

- They perform to a level that would enable any informed observer to identify them as better than comparable groups (ie they clearly 'shine').

- They achieve results with fewer resources than would be assumed to be necessary (ie they are efficient).

- They embody the best values of the organisation culture and act as role models (or reference groups) for the culture.

3.3 Barriers to effective team functioning

There are a number of potential barriers.

- Group norms, unchallenged by the group or group leader, undermining performance (eg restricting output or resisting leadership).

- Lack of support, information or resources from management to fulfil the task (including a lack of genuine decision-making authority and accountability).

- Lack of communication, causing unco-ordinated effort, the potential for misunderstanding, and a lack of shared identity and sense of collaboration.

- Unmanaged conflicts of interest, interpersonal hostility or status barriers blocking team development, co-operation and information-sharing.

- Under-performing or under-motivated individuals holding back the team and causing mistrust and resentment.

- Unchecked team cohesion, diverting attention from the task and causing over-confident decision-making.

- Poor team leadership, creating power conflicts and imbalances, lack of communication and certainty – or (if too rigid) stifling individual creativity and initiative.

- Corporate culture and reward systems which value the individual over the group, creating a disincentive to invest energy in team working.

- Powerlessness – where a team lacks the authority or influence to get its requests met or its recommendations implemented – creating a disincentive to invest energy in the team.

3.4 Elements in team management

A useful overview of team management was provided by Charles Handy (1993), who took a contingency approach to team effectiveness.

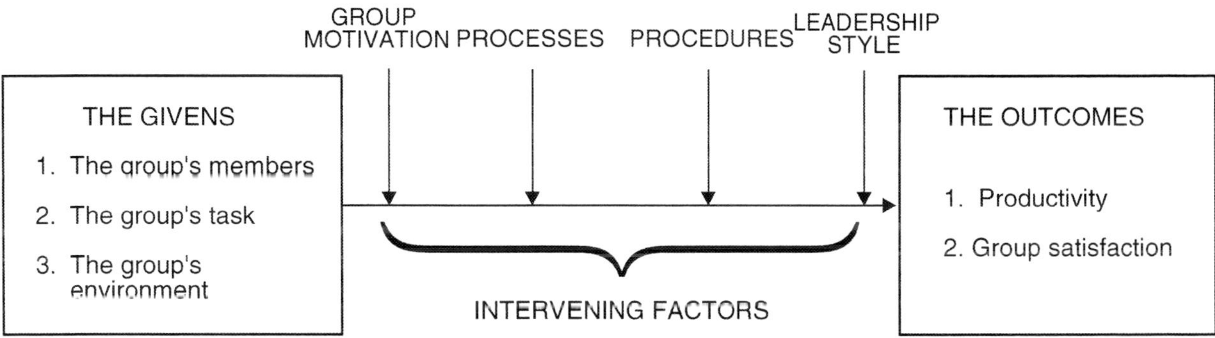

3.4.1 The givens

Some elements of team working may be established by existing arrangements, organisational infrastructure and task structure: they are the raw materials that the leader 'has to work with'.

There may be an established team, providing a mix and balance of skills, competences and personality/role preferences.

The nature and structure of the task will shape the design and allocation of work for the team; its performance goals and standards; its motivation and culture of work; the type of leadership required; the use of technology – and a number of other factors.

The environment includes the physical surroundings and conditions of the team's work but also the place of the team within overall business processes and structures; the constraint of corporate policies and systems the influence of organisation culture and so on.

3.4.2 Intervening factors

Intervening factors are the elements which are potentially within the manager's control, and can be manipulated to bring about positive outcomes for and through the team.

The motivation of individuals, and the team as a whole, is a key influence on team effectiveness.

The task and the organisational authority to enforce compliance with task directives, are givens – but motivation is required in order to get team members to comply and perform willingly and to the best of their ability.

Processes describe how a team performs its functions and maintains steady and constructive team working: how it establishes ways of operating; how it makes decisions; how it structures communication and so on.

Procedures describe the processes laid down by the organisation and team manager for performing tasks: how things should be done in order to achieve outcomes and targets efficiently and effectively. They often take the form of standardised sequences of tasks, and the instructions, tools and guidelines that will support team members in performing them: how to select a venue for a marketing event; how to co-ordinate a promotional campaign; how to establish marketing budgets etc.

Leadership style is, in effect, the way in which all of the intervening factors are mediated to the team: how they are communicated; how influence is used to secure team member 'buy-in' and commitment.

3.5 Managing team processes

3.5.1 Team formation and development

Some teams may be set up from scratch with members who have not worked together before. A team can change its character and dynamics as its members change over time. Project teams are constantly being formed, disbanded and re-formed. It is important to realise that effective teams aren't simply 'put together', they have to develop, and It can take time for a new or altered group to become a fully functioning, performing team.

Four stages in the process of team development were identified by Tuckman (1965).

- **Forming**

    The team is just coming together. Each member wishes to impress his or her personality on the group. The individuals will be trying to find out about each other and about the aims and methods of the team (as set by organisational traditions and standards). There will at this stage probably be a wariness about introducing new ideas, in order to avoid 'rocking the boat'.

- **Storming**

    This frequently involves more or less open conflict between team members, as individuals seek to assert themselves. There may be challenges to the original objectives, procedures and norms established for the group. If the team is developing successfully this may be a fruitful phase, as more realistic targets are set, trust between the group members increases.

- **Norming**

  A period of settling down as a group. Norms and processes will evolve for group functioning: work sharing, output levels, interactions, decision-making and so on. Group cohesion increases as roles (including leadership) are established.

- **Performing**

  The team is functioning well as a group with each member fulfilling clear roles: focus and energy shifts to task performance. Team processes support performance through genuine collaboration (working together).

Tuckman and Jensen (1977) later added two stages to Tuckman's model.

- **Dorming**

  Once a group has been performing well for some time, it may get complacent, and fall back into self-maintenance and member-satisfaction functions, at the expense of the task.

- **Mourning/adjourning**

  The group sees itself as having fulfilled its purpose – or, if it is a temporary project group, is due to physically disband. This is a stage of confusion, sadness and anxiety as the group breaks up. There is evaluation of its achievements and gradual withdrawal of group members. If the group is to continue, going on to a new task, there will be a re-negotiation of aims and roles: a return to the forming stage.

Activity 3

Read the following descriptions of team behaviour and decide to which category they belong (forming, storming, norming, performing, dorming, adjourning).

(a)   Two of the group are arguing as to whose idea is best
(b)   Progress has become static
(c)   Desired outputs are being achieved
(d)   A shy member of the group is not participating fully
(e)   Activities are being allocated by the team

3.6 Team building

Team building is activity by a manager or leader to support, facilitate and, if necessary, accelerate the process of team development. You might approach this by considering what a manager can do to facilitate each of the stages of group development identified by Tuckman (1965).

Some ideas include:

- **Forming**: 'ice breaking' activities; social and 'team bonding' activities to accelerate familiarity and trust; clear articulation of goals and expectations (where possible, clearly stated as being open to negotiation and question by the team).

- **Storming**: encouraging questioning, challenge and constructive criticism; laying down ground rules to keep conflict/criticism constructive and task focused; managing potentially negative conflict.

- **Norming**: team building activity to encourage cohesion and team spirit (discussed further below); providing information and communication mechanisms for decision-making; reinforcing/rewarding positive group norms and challenging negative ones; 'contracting' with the team on agreed ideas, methods and roles.

- **Performing**: backing-off; reinforcing collaborative behaviours; providing support and information where required; monitoring and managing performance.

- **Dorming**: bringing issue into the open; challenging complacency; introducing competition or other incentives to renew motivation.

- **Mourning/adjourning**: bringing confusion, sadness into the open for discussion; encouraging positive evaluation of the team's achievements; motivating and encouraging team members to keep performing to the end; planning handover of activity to the new team (if any); generating positive anticipation of new team and tasks.

3.7 Team identity

More generally, team-building activity is regarded as focusing on three aspects of team working: team identify, solidarity and cohesion.

Definition

Team identity refers to getting people to see themselves as part of the group and to identify themselves with its objectives.

This entails:

- Communicating team identity (eg giving the team a title or nickname)

- Creating a team 'space' (ie a personalised meeting area or intranet page, perhaps)

- Expressing the team's self-image, eg through shared symbols (a 'badge' or 'mascot')

- Mottos and value statements; mythology (sharing stories of the team's successes, heroic failures and funny shared moments)

3.8 Team solidarity

Definition

Team solidarity refers to encouraging cohesion and mutual loyalty, so that members put in extra effort for the team.

This can be achieved by:

- Expressing solidarity (eg using 'we' language)

- Providing opportunities for relationships to develop (eg social interaction time, bonding activities)

- Managing conflict within the team: ensuring that disagreements are openly expressed and resolved

- Controlling competition within the team (eg by avoiding favouritism)

- Injecting an element of competition with other teams (eg competitive awards for best results or productivity)

Definition

Shared objectives: encouraging the team to commit itself to shared work objectives and to co-operate willingly and effectively in achieving them.

This can be achieved by:

- Clearly articulating team objectives and their place in the big picture of marketing, corporate and value chain activity

- Involving the team in setting specific targets and standards and agreeing methods of organising work

- Giving regular feedback on progress and results via team briefings

- Inviting team self-appraisal, feedback and suggestions, so that they can drive improvement planning

- Positively reinforcing behaviour that demonstrates collaboration and commitment to the task (through rewards, recognition etc)

- Celebrating the success of the team and championing it to the rest of the organisation

3.9 Team cohesion

Co-operative groups have been shown to be more effective than competitive groups, where individuals focus on their own contributions rather than the group's shared performance.

Team cohesion or solidarity is broadly regarded as desirable in order to create committed, co-operative working, mutual loyalty and accountability and open information sharing – all of which may help to maximise the potential synergy of team working and individual satisfaction.

However, you should be aware that it is possible for groups to become too cohesive.

Problems may arise in very close-knit groups because:

- The group's energies may be focused on its own maintenance and relationships, instead of on the task. Handy (1993) argued that: 'ultra cohesive groups can be dangerous because in the organisational context the group must serve the organisation, not itself'.

- The group may be suspicious or dismissive of outsiders, and may reject any contradictory information or criticism they supply; the group may be blinkered and stick to its own views, no matter what; cohesive groups thus often get the impression that they are infallible: they can't be wrong – and therefore can't learn from their mistakes.

- The group may squash any dissent or opinions that might rock the boat. Close-knit groups tend to preserve a consensus – falsely, if required – and to take risky decisions, because they have suppressed alternative facts and viewpoints.

- Janis (1972) identified this as groupthink: *'the psychological drive for consensus at any cost, that suppresses dissent and appraisal of alternatives in cohesive decision-making groups'*.

Since by definition a group suffering from groupthink is highly resistant to criticism, recognition of failure and unpalatable information, it is not easy to break such a group out of its vicious circle.

The leader must challenge and encourage the group to:

- Actively seek outside ideas and feedback
- Welcome self-criticism and dissent
- Deliberately evaluate conflicting evidence and opinions, looking for useful learning.

3.10 Group decision-making

Decision-making is a key team process, because:

- Pooling skills, information and ideas from different individuals (or functions, specialisms and levels in the organisation) may increase the quality of the decision.

- Participation in the decision-making process makes the decision acceptable to the group, whether because it represents a compromise or consensus of their views, or merely because they have been consulted and given a sense of influencing the decision.

Team decisions may be arrived at in various ways.

- **Imposition or application of authority by the leader**

    This may be most appropriate where the speed or 'rightness' of the decision are more important than team agreement. The team may be involved in areas such as problem definition and the formulation of alternative solutions, but the leader will take the final decision.

- **Majority rule**

    By vote, or the leader's 'getting a sense' of the view supported by the majority of team members. This may be most appropriate where the acceptance of the group is the most important factor, rather than the 'rightness' of the decision.

- **Consensus**

    A process by which a range of potentially divergent views is examined and persuasive arguments used until there is broad agreement among all members. This takes longer, but is often more effective at the implementation stage, since all members of the group are able to 'own' the decision.

In less effectively functioning teams, this may be a negative process, where decisions are taken without input from team members; by a minority (eg a dominant clique within the team); by default (eg if the leader has abdicated responsibility for decision-making); or not at all (eg the issue is left open until events or higher authority make the decision for the team).

Group decision-making tends to take longer (especially through consensus-seeking), but decisions are often better evaluated and more representative (owing to the input of different viewpoints) and therefore implemented with more commitment.

Key tasks of the leader are to:

- Elicit and utilise team members' views, knowledge and expertise, without abdicating responsibility for the final decision – but without giving the impression that contribution is a waste of time, or that participation is not genuine.

- Encourage progressively less leader-centred decision-making, as the group matures.

- Providing information, coaching and other forms of facilitation and support for group decision-making.

- Avoid the 'risky-shift' phenomenon, whereby groups tend to take greater risks than the same individuals would on their own. There should be rigorous insistence on hearing divergent viewpoints and evaluating options.

3.11 Team communication

Effectively functioning groups tend to move from a leader-centred, leader-initiated pattern of communication to one where interaction is multi-directional or 'all-channel': any member can communicate directly with any other member.

Cliques and isolated individuals become included within this web of interaction over time, facilitating the involvement and contribution of every-member.

Features of effective group communication, to be modelled, coached and supported by the leader, include:

- **Open, honest communication**

    Including the ability to deal with conflicts, issues and criticism openly, directly and fairly (without personal animosity or grudge-holding).

- **Task-relevant information sharing**

  No withholding on a 'need to know' or 'knowledge is power' basis..

- **All-member participation** in meetings, discussions and decision-making. Equitable participation does not mean that all members will share equally, but that all members can get a fair hearing when they have something to say.

- **Absence of artificial status barriers**

  So that senior and junior members communicate with ease.

- **Positive contributions**

  Giving and seeking information, suggestion and opinions; encouraging and affirming others; being appropriately vulnerable; checking understanding; giving constructive feedback; summarising; explaining and so on) outweigh negative contributions (attacking; being defensive; difficulty stating; fault finding; interrupting; overriding others).

Rackham & Morgan (1977) suggest the technique of contribution profiling: analysing the number of contributions of different (positive and negative) types made by each member of a team during a meeting or discussion, to identify dysfunctional behaviours which can be fed back to the individual for adjustment, or countered in communication situations by positive leadership intervention.

Activity 4

Consider what determines the dynamics within a group to which you belong in your organisation, and evaluate how significant they are to the performance of the activities that the group undertakes?

---

4 Background to motivation

Definition

Motives are intrinsic needs and desires existing within us which act as driving forces to search for satisfaction, contentment, a sense of well-being and happiness. Internal drivers that energise behaviour to satisfy a need.

Incentives are extrinsic objects, goals, satisfactions or circumstance which the individual feels will satisfy a need and are inducements to respond. Incentives can be financial or material, for example a holiday or some gift or competition, that are seen as of value as a reward including the admiration of others, or fear of disapproval.

Disincentives are extrinsic circumstances or behaviour that may cause individuals to lose interest, for example lack of recognition of good work, long unpaid overtime or poor working conditions and this may result in a person possibly rebelling quietly or openly.

Intrinsic and extrinsic concepts can be used with effect in designing motivational practice within organisations. For example, intrinsic motivation can provide organisations with lower costs and higher quality together with a reduced need for close supervision. This, in itself, decreaces the reliance on extrinsic motivators alone.

For example, money is often cited as a motivator and while it can have a short-term effect it is an external incentive. Problems arise if incentives are confused with people's motives for behaviour. If we rely on money to motivate it can only do so for a very short time and then people come to expect that money. The only way to motivate again is more money which may not be available. It also assumes that people only work for money. While money is a necessity many of us work for a variety of other reasons and many, of course, work willingly for no money at all.

---

5 Content theories of motivation

Definition

Content theories emphasise particular aspects of an individual's needs or the goals that they seek to achieve as the basis for motivated behaviour.

We now undertake a review of the major content theories.

5.1 Maslow's hierarchy of needs theory

Maslow's hierarchy of needs theory (1943) was developed to understand motivation purporting that individual needs are hierarchical with the stronger, more basic needs at the bottom and weaker more complex needs higher up.

The basis of the model is that individuals have innate needs or wants which they will seek to satisfy which have an in-built prioritising system.

1   **Physiological**

    Needs are the basics: food, drink, sleep, sex, shelter from the elements.

2   **Security/safety**

    Needs are concerned with danger, deprivation, the ability to work and provide security to dependents.

3   **Social**

    Needs concern relationships with others, how a person interacts with others, whether or not they belong, are part of a group.

4   **Self-esteem**

    Needs are about how one is regarded internally to self and externally to others including respect from others, recognition and status and self confidence, independence, achievement.

5   **Self-fulfilment**

    Needs are about personal development and achieving one's own potential to the full in ways far broader than work is just is part of life.

The assumption is that primary needs must first be met before a person can move up to satisfying secondary needs. Moving up to higher order needs does not, however, mean a person will not, at some time, find they again require to satisfy lower order needs.

In the workplace if a job is threatened or change proposed it can often lead people to abandoning weaker needs and refocusing on primary needs.

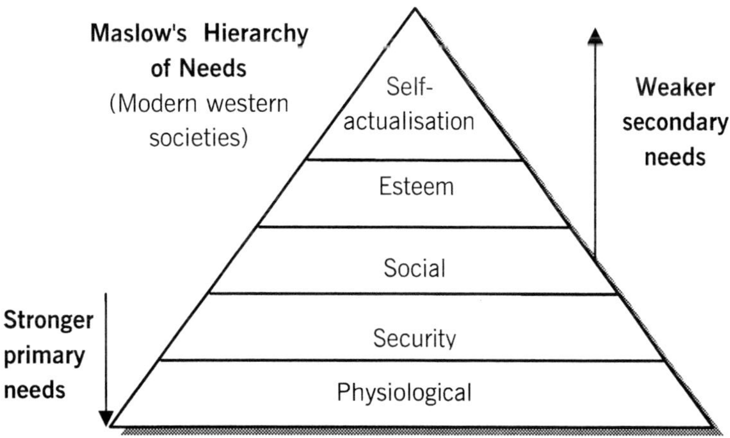

Maslow suggests that the hierarchy displays the following features:

- A need once satisfied is no longer a motivator
- A need cannot be effective as a motivator until those before it have been satisfied
- If deprived of the source of satisfaction from a lower order need it will again become a motivator
- There is an innate desire to work up the hierarchy
- Self-fulfilment is not like other needs; the opportunities presented by it cannot be exhausted

Leaders need to be aware that according to this model once a need is satisfied it no longer motivates. A need satisfied results in other needs emerging and now these needs are the source of motivation. The challenge for the leader is recognising when needs are met, when other needs emerge and how to design work to motivate.

However, critics point to a number of difficulties in applying this model rigidly to the organisation:

- Not everyone is motivated only by things that go on inside the organisation
- People at different stages of their lives will be motivated by different things
- Theory cannot explain all behaviour
- It is a USA theory developed during the 1940s that cannot apply to organisations in today's world
- Organisational events can impact in satisfaction at more than one level in the hierarchy
- Individuals place different values on each need

The value of Maslow's model is that it forces managers and leaders to examine what motivates the employee from the employee's perspective and to seek to understand how they perceive the situation in order to deploy relevant programmes of activity and support.

Global case study

Employees at RBS enjoy Total Reward – a specific benefits package designed by RBS that goes far beyond salary. It offers benefits for each member of staff that include not just money, but also personal choice in working hours and security. The RBS Total Reward package also offers flexible pension funding, health and medical benefits, paid holidays, and a confidential advice service. Employees have a generous holiday allowance (between 25 and 30 days for full-time staff), with the option of buying or even selling days. Employees may also choose from a wide range of lifestyle benefits, including discounted shopping vouchers, childcare facilities and RBS financial products, such as mortgages, currency exchange, personal loans and banking at special staff discounted rates.

http://www.thetimes100.co.uk/case-study--motivating-through-total-reward--106-258-4.php

Accessed 9 April 2010

Suggest ways in which rewards or incentives could be offered to team members to satisfy each of the needs identified by Maslow.

5.2 McClelland's acquired needs theory

McClelland's (1961) theory identified that strong needs are not inherited but are learnt and specified three kinds of needs:

- **Need for achievement**

  To be able to effect desirable outcomes.

- **Need for power**

  To be able to influence and control events.

- **Need for affiliation**

  To connect and associate with others that leads to belonging and relationships.

Entrepreneurs are often people with a high achievement need characterised by a preference for more difficult tasks; personal responsibility for performance; clear and unambiguous feedback and they usually are more innovative. They may enjoy teamwork or be happy to work in isolation depending on the individual.

Power motivates those who have a strong desire to have an impact and be influential which tends to be more important than getting things done. They do not necessarily have to be highly visible or seek the limelight, they are not interested in personal power. What is important is their need for impact and influence and each will interpret this differently. The very nature of how these people are motivated means interaction with others is important but usually as a means to a group or organisation goal. Their need to belong will, again, depend on the individual but it will not be a primary need.

People who prioritise relationships and belonging have a high affiliation need and will be motivated by teamwork. Working on tasks that are designed for individuals with little interaction with others will quickly see them feeling isolated and demotivated.

Satisfaction comes from working with others to achieve task goals and in sharing success.

5.3 Herzberg's two-factor theory

Frederick Herzberg's two-factor theory (1966) undertook comprehensive research in organisations to discover why people enjoyed some aspects of work and found others dissatisfying. Of the factors that emerged as influencing motivation. Herzberg identified what he called motivators (or satisfiers) and hygiene (or maintenance) factors.

Hygiene factors were lower order needs such as the relationship to the environment and elements that if absent, could cause dissatisfaction. Improving hygiene factors reduces dissatisfaction or possibly provides a short lived sense of satisfaction. They include salary, status, working conditions, supervision and job security.

Motivating factors are intrinsic higher order needs that have the potential to provide sustainable satisfaction to individuals and improve performance. They are concerned with the content of the work, together with the way in which it forms a meaningful whole and factors include achievement, recognition, responsibility, growth, advancement and job interest.

If hygiene factors are causing dissatisfaction, for example through low salary or poor social conditions, motivators such as recognition or responsibility will have limited, if any success.

The significance of Herzberg's model is that:

- Lack of positive levels in the hygiene factors does not lead to demotivation, but to dissatisfaction.
- High levels in hygiene factors does not lead to motivation but to non-dissatisfaction.
- High levels in motivation factors leads to positive motivation.
- Low levels of motivating influences reduces overall levels of motivation but does not create dissatisfaction – it creates feelings of non-satisfaction.

The key value of this model is that it reminds leaders that in designing the organisation they need to concentrate on two sets of factors if both motivation and satisfaction are to be maintained.

However critics of the model point to the following:

- Herzberg's results are capable of different interpretations
- The model does not provide for individual differences
- Application of the model can only best be applied to manual and unskilled workers

6 Process theories of motivation

Definition

Process theories provide a model of the interactions between the variables involved in the motivation process.

We now undertake a review of the major process theories.

6.1 The Vroom expectancy model

Definition

Also referred to as a **path-goal theory** it is a model suggesting that it is possible to identify a motivational sequence leading to the achievement of particular goals.

Expectancy theory proposes that people are motivated by different needs depending on the value they attribute to different outcomes and rewards, and that different people have different expectations.

For example, if you wanted and were offered the opportunity by your employer to sponsor you on The Global Marketer Programme in return for higher performance, you would be motivated. However, you would not be motivated if this was not important to you or if you felt that your employer was not able to deliver their promise.

Victor Vroom (1960) was the first person to link expectancy theory to work motivation. He worked out a formula by which human motivation could be assessed and measured, based on three variables:

- **Valence**

    Reward value, a person's feeling about specific outcomes, the value placed on it, a preference for, or the attractiveness of, a particular outcome to the individual.

- **Expectancy**

    That if effort is made it will result in performance (eg quality, quantity).

- **Instrumentality**

    Links together the ideas of the first and second level outcomes.

Expectancy theory can be used to measure the likely strength of an employee's motivation to act in a desired way in response to a range of different rewards, to find the most effective motivational strategy. In particular, it suggests that:

- Intended results should be made clear, so that the individual can gauge the task by knowing what is expected, the reward, and how much effort it will take.

- Immediate and on-going performance feedback should be given. Without knowledge of actual results, there is no check that extra effort was justified (or will be justified in future).

- Individuals should be offered rewards and incentives that they value: this may require some flexibility (eg a 'menu' approach to rewards and benefits) on the part of the organisation.

Managers need to follow through with promised rewards for high performance, in order to increase expectancy for future incentives.

Porter and Lawler (1968) developed the model further by attempting to link motivation and performance. They drew attention to the fact that a range of variances such as an individual's view of work as well as motivation produces performance.

According to Vroom's model people will only put effort into doing a task if the expected outcome or reward is seen as achievable and desirable. The value placed on needs will be influenced by whether they are internal higher order needs such as self-esteem or self-fulfilment which are decided by the individual, and/or external lower order needs given by others such as recognition or promotion.

This requires leaders to relate what people are doing to achieving an outcome and a reward that they will value. It also implies that different people will be motivated in different ways and this is a complex task for leaders and managers.

6.2 Locke's goal theory

Locke (1968) suggested that a person's objectives play an important part in formulating their behaviour. In an organisation this can be used as a mechanism to motivate people to deliver desired behaviours.

For the process to work it is necessary that organisations are able to deliver the objectives that employees desire. For example, it is used as the basis of performance appraisal systems that are a formal feedback mechanism to direct employee behaviour towards the achievement of management objectives.

Within the model the outcome can be significantly influenced by:

- The more specific the goal the more likely it is to be achieved
- The completion requirement (finish date) should be specific
- Goals that are difficult to achieve are more likely to be fulfilled than easy ones

As stated above, this goal-setting approach or motivation is widely used as the basis for performance appraisal systems, particularly where projects, results or change are features of the job. However, given the turbulent nature of the business environment it is increasingly difficult for individuals to maintain performance targeted at specific goals over an extended period.

7 Additional perspectives on motivation

7.1 McGregor's theory x and theory y

McGregor (1957) goes to the heart of people's assumptions about each other that can still be seen in the workplace today. McGregor argued managers' attitudes to, and assumptions about, employees lead them to manage in a particular way and offered two opposing views on these assumptions.

Theory X assumptions

- The average human being has an inherent dislike of work and will avoid it if they can.

- People work mostly for money, status and rewards.

- Because of the characteristic of dislike for work most people must be coerced, controlled, directed and/or threatened to get them to put adequate effort into achieving organisation goals.

- The average human being prefers to be directed, wishes to avoid responsibility, has relatively little ambition and wants security above all else.

Such assumptions are likely to lead to a more directive and autocratic, even bullying, style of leadership.

The assumption is that people will not take initiative and the best that can be achieved is to push people hard and often only provide money as a reward.

This neglects significant drivers for performance.

Theory Y assumptions

- The expenditure of physical and mental effort in work is as natural as play or rest.

- External control and the threat of reprisals are not the only means of encouraging commitment to organisation objectives. People exercise self-direction and control in the service of objectives and to commit to objectives is a function of the rewards associated with their achievement.

- The average human learns under proper conditions to accept and also to seek responsibility.

- The capacity to exercise a high degree of imagination, ingenuity and creativity in the solution of organisational problems is widely, not narrowly, distributed in the population.

- Intellectual potential of the average person is only partially utilised in the conditions of modern life.

Theory Y assumptions lead to a more democratic style of leadership and tapping into people's desire for work and for achievement is the route to success. If the right conditions are created people will take the initiative, be prepared to solve problems and can be motivated by a range of factors other than money.

7.2 The challenge of management

Managers are faced with major difficulties in attempting to motivate employees and a range of theories to choose from. Whilst each has something to offer they largely emanate from a western perspective and may not be valid in other cultural settings, particularly given the rising importance of emerging economies.

The wide range of motivational strategies and activities that can be deployed make it difficult to optimise motivational strategies and maximise performance. Additional challenges are in seeking to motivate employees across generations, countries and time zones.

Discussion

Study buddy

Select any one example of how you (or your own manager) approach motivation in your organisation.

- What approach is being used? What (intrinsic or extrinsic) rewards or incentives are offered? How clear are the attainment targets by which staff will earn those rewards? What kind of feedback on progress/performance is given?

- How effective is this approach? Are team members highly motivated? Do they perform up to expectation or beyond? Are they satisfied in their work?

- Use any one motivation model (eg Herzberg or Vroom) to explain the effect of the approach used on the motivation and performance of the team? If there are motivational problems in the team, what means of improving the manager's approach might be suggested by this model?

Discuss your findings with your Study buddy.

8 Authority, power and influence

Authority, power and influence while similar are not the same but are often used together to initiate action and achieve goals.

Definitions are many and varied and can be confusing but Vecchio (2007) provides a useful clarification.

> **Definition**
>
> **Authority** infers a notion of legitimacy usually from status or position and is therefore seen as a right to try to change or direct others. Authority is dependent on – and often constrained by – position.
>
> **Power** is the ability to change the behaviour of others and perform actions that they might not otherwise undertake. Power is dependent on particular sources.
>
> **Influence** tends to be subtler, broader and more general than power and while it also is the ability to change behaviour it is less reliable and weaker than power.

Influence is dependent on particular tactics. Organisation culture and leadership style strongly affect the nature, sources and use of authority, power and influence in an organisation. If the organisation is hierarchical, the culture subservient and leadership style autocratic, the sources of power may be from formal status or seniority and positions held.

Power will typically be used to dictate and control. In some organisations, particularly where a high level of expertise is required, power may reside, for example with technicians or scientists, by virtue of their expertise.

Employees have become more powerful partly as a result of changing attitudes and employment law but also because in many industries high levels of skill and/or knowledge are required and thus employees have power by virtue of that skill or knowledge.

8.1 Sources of organisational power

It might be assumed that a leadership position automatically leads to a position of power. However, power does not inevitably come with leadership. Power comes with the territory but it is worth considering the sources of this power.

Different types of power have been identified by French and Raven (1959).

Type of power	Characteristics
Legitimate power	Comes with authority as it is based on holding a formal position. People who accept this legitimate power will comply and actively seek direction and guidance from such a source of power.
Reward power	Based on giving something of value to someone in exchange for particular behaviour. These rewards may be financial, for example a bonus, or non-financial for example an award for achievement.
Coercive power	Stems from the threat of punishment, the most obvious being no salary increase or bonus, no promotion and dismissal. Less obvious, and potentially more damaging, are bullying and social exclusion. In the real world we know there are some people who just do not want to work or work well for a variety of reasons. However, coercive power, while suggested by some as an acceptable form of power, is a last resort.
	The use of coercive power can suggest a failure in leadership and people management. This does not mean we should tolerate poor performance or paralyse ourselves with political correctness. If a person is not responding to every opportunity to improve, disciplinary action or dismissal are necessary.

Type of power	Characteristics
Expert power	Comes from knowledge and experience and being seen as talented, gifted, the best in the business. It is usually narrow in its ability to exert power constrained by the nature of the expertise.
Referent power	Comes from people with, for example special qualities, poise, great personality, ability to inspire and interpersonal skills, charisma. People acquiesce to this power because they admire, respect or identify with the qualities this person has.

A leader can possess and utilise a number or all of these sources of power. A leader can hire and fire, dictate terms and decide on rewards, particularly if they own the company.

A leader can impress and reassure people with their advice and mesmerise. This power may only achieve obedience, compliance and even loyalty. But today's leaders have to achieve so much more.

One of the greatest influences on leadership style evolution today is the recognition that among other goals leadership style needs to inspire and empower people; build employee commitment and loyalty; nurture an ethical culture of creativity for innovation to improve organisation performance through excellent people performance.

These goals require a leader to build credibility, respect and trust. And this is where real power comes from. It is useful to understand sources of power and to utilise them.

8.2 Politics and power

Politics is part of human nature and leaders need to use and cultivate the more positive and productive forms of politicking but be very clear negative politicking is unacceptable. The many forms it takes include:

Type of politic	Positive	Negative
Networking and coalitions	Building contacts and nurturing relationships across the organisation and externally to advance group and organisation goals	Identifying potential competitors (individuals) and allies to recruit in order to destroy those seen as being in competition
Ingratiation		Using compliments, flattery, agreeing with someone considered important regardless of the right or wrong or truth, are all designed to create a favourable impression and win approval and support
Personal promotion	Establishing credibility to sway and influence decisions in order to secure desired outcomes for the organisation or group	Egotistical promoting of often exaggerated achievements no matter how small or relevant for reasons such as self-importance
Information	Used to convince, persuade and provide evidence for a particular argument or justify a particular point of view	"Information is power" syndrome and information can be used in the other forms of politics mentioned
Upstaging		Publicly criticising a subordinate, peer or manager, bragging about "what you have done" or "how you would have done it", taking credit for the ideas and achievements of others

Usually the tactics employed, while having some self-interest, often have the interest of a task, project or department etc, as the goal. In these circumstances people find ways to get what they think are the right things done. They may be competing for resources or a project that is not necessarily a priority for others, particularly those in control.

Politics becomes negative when all or some of these tactics are used purely for self-interest and often in conflict with the goals of others and the organisation. When political tactics cause others to be se en in a poor light or it damages performance it can be a destructive force in the organisation.

How often have you wondered or heard others say they do not know how a decision got made as it seems irrational, did not appear to go through the formal decision-making process or channel etc. Politics will often be the reason behind what appears to be irrational decisions and leaders should never underestimate the power of politics in the informal decision process.

This process includes many more people than a formal decision process and they are not always easy to identify. One of the reasons people engage in politics is because people have different perspectives, agendas and interpret needs and decisions differently. This can bring people into conflict and result in negative politicking.

Global case study

What is one to do knowing that politics is widely at work in the world? You have two choices in my experience. You have to play the office politics game if you want to win. Don't fight the idea instead embrace and enjoy it. For most people, it is hard to do something you don't like. And it can be emotionally draining pretending to like someone you don't. Alternately, you could be so good at your job that the company cannot afford to lose you. This demands nearly the same effort as faking company/co-worker relationships with supervisors. Doing one of the two is the only way to respond to politics at work. I advise the second, performing your job so well that politics at work can't impact you. You obtain so much influence at the office that you can influence outcomes, regardless of what others say. The aspect could to your surprise lead you to your goal. Determining if you want to play the game, is an important task of making politics at work. In order to achieve success in the realm of business, you need to partake of the game of politics at work. You will be able to determine which path to follow. Results always override office politics.

http://www.articlefriendly.net/Art/31108/340/Workplace-Politics-How-Do-You-Play-The-Game.html

Accessed 11 April 2010

8.3 Influence

Leaders with authority and power still need to be able to influence, particularly today in organisations where employees are technically skilled or are knowledge workers or in organisations where equality is emphasised or workers are volunteers.

Influence often results in people conforming to particular decisions and behaviour. Herbert Kelman (1974) identifies three major types of social influence.

1   **Compliance**

    - People will agree with each other to ensure they receive a positive response and avoid unfavourable reactions.

    - Influence is exercised because of people's apprehension about the social consequences of agreeing or not agreeing with others who have the power to reward or punish.

2   **Identification**

    - People will agree with others who they perceive as having qualities they value in order to establish or maintain a fulfilling relationship that re-enforces their self image.

   Aligning the organisation to the vision

- Influence is exercised because of a desire to promote goals associated with forming and maintaining a relationship which is in harmony with their social identity.

3   **Internalisation**

- People will agree with others who they perceive as having attitudes that are credible and believable.

- Influence is exercised because the issues fit broadly with their own values and goals.

Two main types of influence used in organisations:

1   **Informational influence**

- People can be manipulated by providing both accurate and inaccurate information to achieve a desired outcome.

- People look for information when what they have is ambiguous, when they want expert advice or when a problem or crisis occurs.

- If people giving the information believe the information may have a significant impact they are more likely to want to be accurate.

- The less important the outcome, the more likely people are to conform to the group rather than worry about accuracy of the information.

2   **Normative influence**

- This is about being liked therefore people will conform to the behaviour and decisions of the rest of the group.

- They may not necessarily agree with what others are doing or saying but will conform.

- The more people in the group the less influence one person has.

8.4 Use of power and influence

Much of what has been discussed under leadership styles provides clues to how leaders use power and influence and the organisation culture will determine how much people feel they must conform or have the freedom to disagree or be different.

Leaders and employees in organisations often try to influence each other in different ways and for a variety of reasons. Typically we are:

- Convincing others through logical argument

- Persuading others by presenting a compelling case

- Negotiating with others to gain agreement and usually with the best of intentions and the interests of the group or organisation.

Activity 6

Identify sources of power and influence in your organisation and observe how it is used. When is it used positively and when is it used negatively?

1 Compare and contrast groups and teams

- There are two types of groups: informal groups and formal groups.

- People tend to contribute different skills and attributes to the organisation as group members than they do as individuals.

- All teams are groups – but not all groups are, strictly speaking, teams.

- Teams have a number of advantages such as facilitating the performance of tasks, co-ordination of the work of different individuals or groups, interactive communication and so on.

- Teams and team working are very much in fashion, but it is important to recognise their potential drawbacks, limitations and challenges.

2 Evaluate team structures

- There are two basic approaches to the organisation of team work which requires a variety of skills and competences: multidisciplinary working and multi-skilled working.

- Virtual teams bring together individuals working in remote locations, reproducing the social, collaborative and information sharing aspects of team working.

- Self-managed teams are permanent structures in which team members collaboratively decide all the major issues affecting their work.

- Team membership may be dictated by existing arrangements, organisational appointment or election.

- R Meredith Belbin (1993) drew up a widely-used framework for understanding roles within work teams. Effective team working requires a mix and balance of all the roles.

3 Develop an effective team

- The task of the leader is to build a 'successful' or 'effective' team.

- There are a number of factors, both quantitative and qualitative, that may be assessed to decide whether or how far a team is functioning effectively.

- Vaill (1989) identified the characteristics of 'high performance' or extraordinary teams.

- There are a number of potential barriers to high performance such as lack of support, lack of communication, unmanaged conflicts of interest, under-performing or under-motivated individuals and so on.

- A useful overview of team management was provided by Charles Handy (1993), who took a contingency approach to team effectiveness.

- Four stages in the process of team development were identified by Tuckman (1965).

- Team building is activity by a manager or leader to support, facilitate and, if necessary, accelerate the process of team development.

- Team identity, team solidarity, shared objectives and team cohesion all have an impact on the dynamics, effectiveness and decision-making ability

4 Identify the key elements of motivation

- Motives are intrinsic needs and desires existing within us which act as driving forces to search for satisfaction, contentment, a sense of well-being and happiness.

- Intrinsic and extrinsic concepts can be used with effect in designing motivational practice within organisations.

5 Evaluate both content and process theories of motivation

- Content theories emphasise particular aspects of an individual's needs or the goals that they seek to achieve as the basis for motivated behaviour.

- The basis of Maslow's hierarchy of needs theory is that individuals have innate needs or wants which they will seek to satisfy which have an in-built prioritising system.

- The value of Maslow's model is that it forces managers and leaders to examine what motivates the employee from the employee's perspective.

- McClelland's theory believed that strong needs are not inherited but are learnt and identified three kinds of needs.

- Frederick Herzberg undertook comprehensive research in organisations to discover why people enjoyed some aspects of work and found others dissatisfying.

- Process theories provide a model of the interactions between the variables involved in the motivation process.

- Victor Vroom (1964) was the first person to link expectancy theory to work motivation.

- This goal-setting approach or motivation is widely used as the basis for performance appraisal systems.

6 Assess the manager's role in motivation

- McGregor argued managers' attitudes to, and assumptions about, employees lead them to manage in a particular way and offered two opposing views on these assumptions, Theory X and Theory Y.

- Managers are faced with major difficulties and challenges in attempting to motivate employees and a range of theories to choose from.

7 Consider the effects of authority, power and influence

- Authority, power and influence while similar are not the same but are often used together to get action and achieve goals.

- Power comes with the territory but it is worth considering the sources of this power.

- Different types of power have been identified by French and Raven (1959).

- Politics is part of human nature and leaders need to use and cultivate the more positive and productive forms of politicking but be very clear negative politicking is unacceptable.

- Politics becomes negative when all or some of these tactics are used purely for self-interest and often in conflict with the goals of others and the organisation.

- Influence often results in people conforming to particular decisions and behaviour. Herbert Kelman (1974) identifies three major types of social influence.

Activity 1

This will depend on your own organisation. Consider the benefits and drawbacks of using teams and the implications that team working has for achieving your organisation's vision.

Activity 2

(a)   Completer-Finisher
(b)   Implementer/Company worker
(c)   Specialist
(d)   Plant
(e)   Resource investigator
(f)   Shaper
(g)   Team worker
(h)   Co-ordinator/Chairman
(i)   Monitor-Evaluator

Activity 3

Categorising the behaviour of group members in the situations described results in the following:

(a)   storming
(b)   dorming
(c)   performing
(d)   forming
(e)   norming

Activity 4

The answer here will depend on your own organisation. How new or established the group is, the degree of shared values it has, the way in which communication takes place within and outside the group, the way it is perceived amongst peers, its track record of success and the visible support and commitment the group has from senior management within the organisation will all contribute.

Activity 5

- **Self-actualisation**: job challenge, task variety, development opportunities

- **Esteem needs**: merit pay increase, high-status job title, status symbols, recognition from the team leader, weekly/monthly performance awards

- **Social needs**: compatible work group, opportunities for social interaction with workmates, team identity

- **Safety needs**: job security, basic welfare benefits, consistent leadership, maintenance of the *status quo* (or good management of change)

- **Basic/physiological needs**: basic salary, safe working conditions, stress management.

Activity 6

Sources of power will be affected by the organisation structure and culture, the type of industry and, for example the level of expertise required by your industry. Sources may be obvious such as position in the organisation or the esteem and respect with which someone is held or less obvious, for example, someone with strong influential skills. Identifying when it is used positively might include achievement of organisational goals, resources allocated to an essential project and negatively might include goals that

have more to do with personal ambitions than the good of the organisation and others, or slowing down activities through the introduction of unnecessary steps in a process.

Adair, J., (1983). Effective Leadership: A self development manual. London: Gower.

Belbin, R. M., (1993). *Team Roles at Work*. Oxford: Butterworth Heinemann.

French, J. R. and Rowen, B., (1959). 'The bases of social power'. In D. Cartwright (ed) *Studies in Social Power*, P. 150–167. Ann Arbor: University of Michigan.

Handy, C. B., (1993). *Understanding Organisations*, 4th edition. Harmondsworth: Penguin.

Herzberg, F., (1966). *The Motivation to Work*. London: Wiley.

Huczynski, A. & Buchanan, D., (2001). *Organizational Behaviour: An Introductory Text*, 4th edition. Harlow: FT Prentice Hall.

Janis, I. L., (1972). *Victims of Groupthink*. Boston: Houghton Mifflin.

Katzenbach, J. R. & Smith, D. K., (1993). *The Wisdom of Teams: Creating the High Performance Organisation*. Boston: Harvard Business School.

Kelman, H.C., (1974). Social influence and linkages between the individual and the social system: further thoughts on the processes of compliance, identification and internalization. In J. Tedeschi (ed) *Perspectives on Social Power*, p. 125–71. Chicago: Aldine.

Locke, E.A., (1968). 'Toward a theory of task motivation and incentives'. *Organizational Behaviour and Human Performance*, Vol. 13, p. 157–189.

Mallon, M. and Kearney, T., (2001). 'Team development at Fisher and Paykel: The introduction of "Everyday Workplace Teams"'. *Asia Pacific Journal of Human Resources*, Vol. 39, No. 1 (2001), pp. 93–106.

McClelland, D., (1961). *The Achieving Society*. London: Van Nostrand Reinhold.

McGregor, D., (1957). *The Human Side of Enterprise*. Oxford: McGraw Hill.

Maslow, A. H., (1943). *Motivation and Personality*. London: Harper & Row.

Peters, T. J. & Waterman, R. H., (1982). *In Search of Excellence*. New York: Harper Collins.

Porter, L. W. and Lawler, E.E., (1968). *Managerial Attitudes and Performance*. Homewood IL: Irwin.

Rachlan, R. and Morgan, T., (1977). *Behaviour Analysis in Training*. London: McGraw Hill.

Sobel Lojeski, K. and Reilly, R. R., (2008). *Uniting the Virtual Workplace: Transforming Leadership and Innovation in the Globally Integrated Enterprise*. New Jersey: Wiley.

Steiner, I., (1972). *Group Process and Productivity*. New York: Academic Press.

Tuckman, B. W., (1965). 'Developmental sequence in small groups'. *Psychological Bulletin*, Vol. 63, pp. 384-389.

Tuckman, B. W. and Jensen, M. A., (1977). 'Stages of small group development revisited'. *Group and Organisation Studies*, Dec, Vol. 2, p419.

Vaill, P. B., (1989). Managing as a Performing Art. *New Ideas for a World of Chaotic Change*. San Francisco: Jossey-Bass.

Vecchio, R. P., (2007). *Power, Politics and Influence*. Paris: University of Notre Dame Press.

Vroom, V., (1960). *Work and Motivation*. London: Wiley.

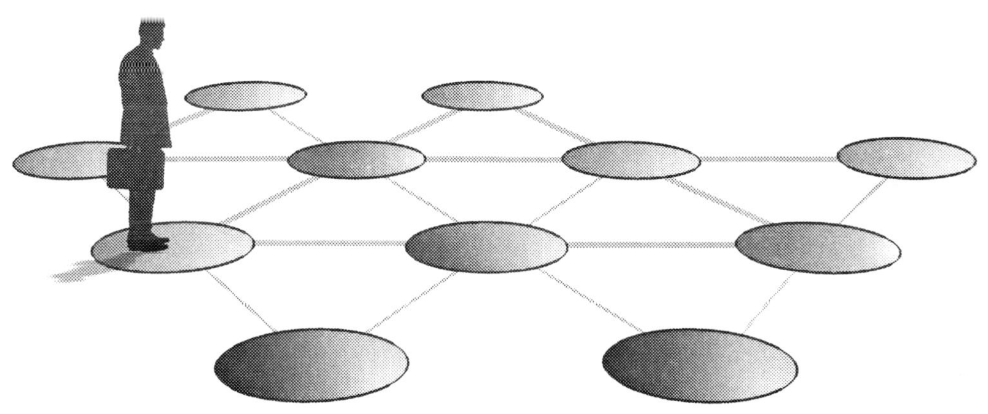

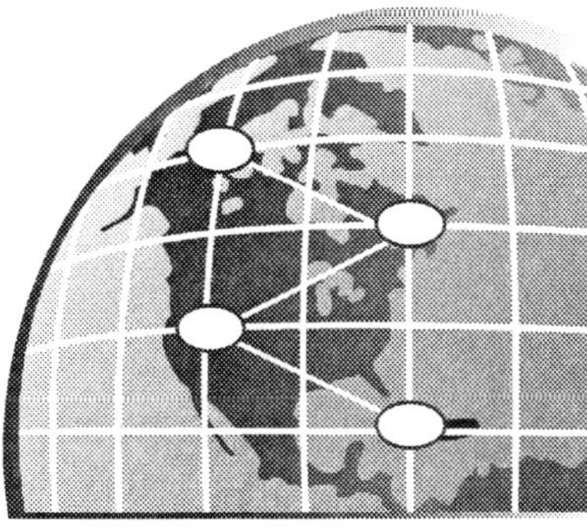

3 | chapter

Leading the organisation to create a culture of innovation and entrepreneurship

We live and work in cultures most of the time not thinking much about what the cultures are or their influences on us. The task of developing and maintaining an innovative and entrepreneurial culture in a long established and large organisation is always going to be more difficult than a smaller organisation. Leaders have the potential to have the most powerful influence on the organisation and its structure, operations, culture, people, competence and performance. They do not do this alone and require management, employees and other stakeholders to influence the shape and form of the organisation. A strong leader can have the most direct influence in shaping, for example a creative and innovative environment, a learning organisation, strategy formulation, collaboration and alliances and stakeholder relationships. The larger the organisation gets the more challenging it is for leaders to create the sort of organisation they want. Knowing what they should do is a small part of achieving their goals and the vision. Knowing what to do is a good starting point and if the right objectives and measures are put in place then leaders can monitor their progress and take corrective actions where and when necessary, and the closer they will get to aligning the organisation to the vision.

This chapter looks at the role of leadership and innovation along with thought leadership and knowledge management.

We then move on to consider the complexities of organisational culture and how a marketing-oriented culture may be achieved.

Contents

Chapter learning outcomes

In this chapter you will cover the following:

- Evaluate the role of leadership in encouraging innovation

- Consider the role of thought leadership and knowledge management in organisations

- Identify and explain why collaboration is important and how to manage it effectively

- Explore how stakeholder relationships can be managed

- Identify your desired leadership style

- Define organisational culture

- Evaluate cultural frameworks and models

- Compare organisational orientations

- Critically evaluate the importance of shared values

1 Leadership and innovation

Few people question the value of creativity and innovation, particularly in our rapidly changing world today but organisation success and survival are dependent on innovation.

Leaders and their managers do not deliberately try to inhibit creative and innovative behaviour but for a variety of reasons this is what often happens, even when the opposite is a clearly stated objective.

Reasons include:

- **Risk averse culture**: past failures, costs, fear of getting it wrong and so on

- **Organisation culture**: responses to past failure (eg a "name and blame" culture)

- **Leadership style and practice**: discourages creativity by emphasising and rewarding uniformity, task completion and an unwillingness to allocate appropriate resources

Global case study

McKinsey research revealed that 65% of senior executives they surveyed were disappointed in their ability to stimulate innovation and few confident of the decisions they made in this area. 94% said that people and organisation culture are the most important drivers of innovation. Often the necessary effort put into maximising efficiency of operations results in standardising everything the organisation does using rigid processes and procedures to protect the organisation from deviation and this has the effect of inhibiting creativity. Establishing an environment for innovation benefits from an understanding of what drives innovation and what is required.

Definition

Creativity is the ability to think differently, to generate new ideas and requires freedom, inspiration, originality, ingenuity.

Innovation is the first attempt to put invention (the idea, novelty, something unique) into practice and requires a process.

Not everyone can be innovative or creative but many never have the opportunity to realise their full creative potential. Organisations need creativity that leads to innovation on new ways of doing things and new things to do. Innovation requires a rigorous process to ensure what emerges from creativity is commercially viable. This is how businesses grow and prosper.

1.1 Environmental scanning and different contexts

Ideas come from a network of sources and are often driven by external environmental change. Monitoring the external environment is part of the innovation process and interpreting analysis findings, in the right organisation climate, stimulates creativity.

Creativity responds to changing market dynamics whether it is macro, for example legislation, economic conditions or micro, for example a change in customer or competitor behaviour.

In a creative climate ideas are also created internally, sometimes without outside influence. This is a natural process and should be encouraged. The innovation process will check the viability of internally generated ideas for their suitability and feasibility to the organisation and externally.

No organisation is exempt from the need to innovate. Private companies, public sector organisations, charities, organisations in stable and simple environments, those in dynamic, complex environments and small, medium and large organisations all at some time or another need to innovate what they do and/or how they do it. What differs is the speed of response and the nature and extent of innovation required.

The global business environment continually changes but organisations need to assess the nature of change to decide whether it is appropriate or necessary to respond. Many managers are frustrated by a culture of continuous change that is not always necessary or beneficial. This can lead to following every fad and fashion regardless of its relevance or value. The skill of leadership is to recognise what to change and when.

1.2 Who is responsible for innovation?

Leadership is responsible for creating the vision for the organisation and the right climate for creativity that leads to innovation that will help achieve the vision. Only when the right climate has been created can everyone else take responsibility for creativity and innovation.

Many leaders genuinely believe they encourage creativity and innovation. However, there are a number of obstacles unintentionally created by leaders and actively maintained by management including a lack of resources, a "get it right first time" mentality and time.

1.3 What innovation and where?

Innovation is not confined to new products and services but is required at various times in other areas including:

- **Operations**: technology and systems that produce products or services

- **Processes**: for managing and delivering value to customers

- **Value**: products and services produced for customers and stakeholders

- **Marketing**: research can require creativity when information is difficult to obtain and communications require adoption of different approaches to rise above the "noise" and reach the desired targets

Leaders need to enthuse managers and employees with the task of finding both short-term innovation, often the marketing mix, and longer-term innovation, for example corporate strategy.

1.4 Type of innovation

1.4.1 Incremental innovation

Less risky, less costly and more likely to succeed and the business learns as it evolves. Incremental innovation supports consistency and continuity and usually builds on something customers understand so they can appreciate the "new" benefits.

Revolutionary innovation involves "Megaprojects" or "do everything" and requires heavy investment. The "unknowns" are greater including market response and learning curves are far steeper. It is only usually an option if a matter of survival or the innovation is known to meet a significant identified need and has guaranteed commercial viability.

Revolutionary innovation may not be consistent. A concerted effort, usually by marketing is required to establish or maintain a clear competitive position, and not lose or confuse customers and stakeholders.

All innovation should lead to competitive advantage and opportunities for differentiation. How long term or sustainable this competitive advantage is depends on the nature of innovation.

Teresa M Amabile (1998) suggests there are three components to creativity.

"Within every individual, creativity is a function of three components: expertise, creative-thinking skills, and motivation. Can managers influence these components? The answer is an emphatic yes – for better or for worse – through workplace practices and conditions."

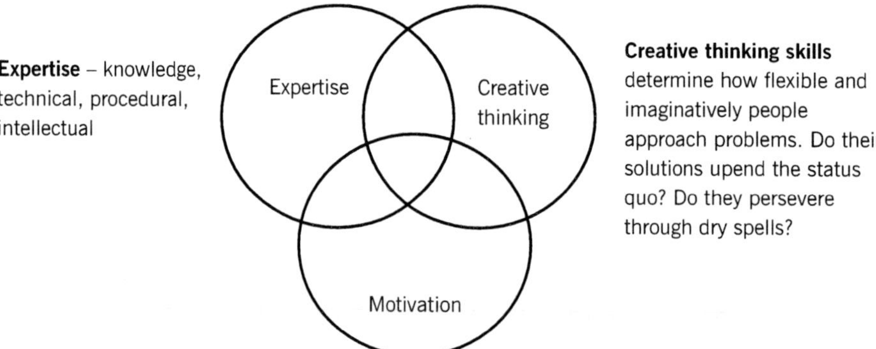

Expertise – knowledge, technical, procedural, intellectual

Creative thinking skills determine how flexible and imaginatively people approach problems. Do their solutions upend the status quo? Do they persevere through dry spells?

Not all **motivation** is created equal. An inner passion to solve the problem at hand leads to solutions far more creative than do external rewards, such as money. This component called intrinsic motivation is the one that can be most immediately influenced by the work environment.

The point made in the diagram about motivation is important. Those people motivated by money to solve a problem will not necessarily be motivated to solve it in a creative or innovative way. It does not automatically follow that it will encourage exploration, experimentation and so on. If money is the reward then solving a problem as quickly as possible, rather than finding the best solution, is likely to be the goal.

Many innovations are improvements on existing ideas or re-designed for new and different purposes and in different contexts. As many trends, preferences and tastes usually change gradually many innovations respond gradually, with modifications and adjustments.

During periods of significant change innovation requires a leap into the unknown and leadership that succeeds in creating and maintaining a creative climate will have built an organisation in a better position to take that leap.

No matter how successful or if the organisation is a market leader leaders can never be complacent. A leader should never lose sight of the need to nurture creativity. This is most successfully achieved through people.

What are the barriers and drivers of creativity and innovation in your organisation? Why do these barriers and drivers exist? Use your knowledge gained from the previous two chapters to provide your own assessment of the causes. Suggest ways to overcome barriers.

In today's innovation-driven economy, understanding how to generate great ideas has become an urgent managerial priority. Harvard Business School professors, Amabile and Khaire, convened a two-day colloquium of leading creativity scholars and executives from companies such as Google, IDEO, Novartis, Intuit, and E Ink. At the event, a new leadership agenda began to take shape, one rooted in the awareness that you can't manage creativity – you can only manage for creativity.

A number of themes emerged: The leader's job is not to be the source of ideas but to encourage and champion ideas. Leaders must tap the imagination of employees at all ranks and ask inspiring questions. They also need to help their organizations incorporate diverse perspectives, which spur creative insights, and facilitate creative collaboration by, for instance, harnessing new technologies. The participants pointed out that they need to clear paths through bureaucracy, weed out weak ideas, and maximize the organization's learning from failure.

http://hbr.org/product/creativity-and-the-role-of-the-leader/an/R0810G-PDF-ENG
Accessed 27 March 2010

2 Thought leadership

It is worth noting there are some different views on what thought leadership is and how it works. Some believe it is about leading an organisation and others that anyone in the organisation can be a thought leader. It is about thinking differently, creating leading edge knowledge and influencing others to think differently, that new ideas are generated and assimilated.

Definition

Thought leadership is shared by all who can think creatively or differently. At any point in a meeting, casual discussion, solving a problem or heated debate different individuals may be demonstrating thought leadership. It is immediate and does not have to have a vision or journey. It is argued that thought leaders are more like artists and therefore do not need to worry about their interpersonal skills or ability to get on with people.

Thought leaders typically include the mavericks and rebels in the organisation who cannot and will not conform.

All organisations need people who can think differently. When it comes to thinking differently one of the most popular and enduring writers on the subject is Edward De Bono (1970). He recognised there are broadly two approaches to thinking as follows:

Vertical thinking	A logical, careful and analytical process using traditional methods to solve problems and find solutions.
Lateral thinking	Adopts an unorthodox approach to solving problems and generating new ideas.

Vertical thinking tends not to create new ideas or different perspectives, it follows trusted paths so comes up with similar ideas and solutions.

De Bono (1970) contrasted this with lateral thinking which requires different ways of looking at things.

Lateral thinking should:

- Generate alternatives, never accept one or two solutions and the more unlikely the alternative the more promising the potential solutions for being different

- Adopt a random approach avoiding following a logical order, instead join up a number of apparently disconnected themes and see what emerges

- Deviate, avoid the obvious path and follow a different one, preferably several different ones

De Bono (1970) believed both vertical and lateral thinking were important, both needed for different purposes and situations.

Robin Ryde (2007) picks up this theme in his book *Thought Leadership*. He shares experiences from running programmes for leaders and potential leaders and from problem-solving workshops.

He describes two dominant patterns in the way people explore issues when they come together.

1 **Deficiencies**

   A tendency to look for the faults and deficiencies in any information they have acquired and in ideas, discussion or possible solutions. Deficient thinking can put a lot of time and energy into proving there are deficiencies. These are the people who see the "problem, difficulty or impossibility" of anything different or new.

2 **Commonsense**

   Thinking applies a broad and general set of principles to a situation to advance the thinking on the subject. As you would expect it is sensible. A problem arises when commonsense can be sensible but not necessarily appropriate.

Sometimes a variety of factors, not all obvious, need also to be considered that might result in a different outcome.

People come to various situations such as solving problems or generating ideas in two ways.

1 With a closed mind, they already have the solution or think there is no need for a solution and are less likely to listen to different views except to criticise or find fault.

2 With an open mind and willing to listen to different points of view, ideas and explore these ideas, they will build, contribute and challenge constructively.

Ryde (2007) describes six standard thinking repertoires that can stifle creativity because they take the most direct route or a path already taken by others.

Deficit thinking	Focuses on faults, shortcomings and weaknesses of the topic.
Rational thinking	Logical, sequential approach to problems or ideas, fails to deal with feelings and emotions typical of business interactions.
Sticky thinking	Forms associations from one person's contribution to the next, conversations wander through association from one subject to another, often avoiding the topic.
Commonsense thinking	Applies general and inexpert knowledge to find a solution not always fit for purpose.
Binary thinking	Assumes solutions are one thing or another, opposing ends of a spectrum, black or white. There are no possibilities in variations or of combinations.
Equity thinking	Uses fairness as its overriding principle, people make comparisons to look for inconsistencies that support the view it is "unfair" and seek to even things up.

Ryde (2007) acknowledges that some of the above "standard thinking repertoires" have their place. For example, challenging and pointing out shortcomings can be important where there are shortcomings. The problem is when the above becomes a habit, particularly just one way of thinking.

Therefore, the first step is to recognise and understand your thinking habits. Do you just have one way of thinking? How rigid is it? What productive outcomes result? Why has this thinking habit emerged? What is the trigger? What factors influence your way of thinking?

The next step is to develop broader thinking skills. This requires discipline and determination, particularly in everyday interactions and under pressure of work. In a meeting or conversation when you recognise you are slipping into your thinking habit try a different approach, deliberately think and say something different.

If you have a tendency to see shortcomings try instead to develop the discussion to possible "what ifs". Ryde recommends developing a broader repertoire by adopting what he calls the "shadow side" of the dominant standard ways of thinking.

Dominant function	Shadow function
Deficit thinking	Strengthen-based thinking
Rational thinking	Feeling thinking
Commonsense thinking	Insight thinking
Equity thinking	360 degree thinking
Binary thinking	Re-integrated thinking
Sticky thinking	Exit thinking

2.1 Shadow function

Strengthen-based thinking	Searching for the positive in a situation and supported by evidence, "can do" mentality.
Feeling thinking	Allows for emotions and feelings in a conversation and is not threatened or embarrassed by them, it is accepted as part of what we are. It does not mean events are overtaken by emotions but rather that we acknowledge likely reactions and emotions and we use good interpersonal skills to manage feelings.
Insight thinking	Drawing on experience and expertise and using this to inform ideas and possible solutions.
360 degree thinking	Requires adopting multiple rather than single perspectives on any given problem. Seeing ideas and problems from many perspectives and trying to understand how these different perspectives change as a result.
Re-integrated thinking	Avoids "either/or" and looks to integrate, combine, adjust where it is appropriate to do so. It requires exploring, investigating and being prepared to work on ideas or solutions with a broader perspective.
Exit thinking	Avoiding slipping into meandering conversations that go off topic and being prepared to refocus. Also being prepared to bring discussions to an end when they cease to add value or contribute something useful.

Changing unproductive thinking habits and developing effective thinking encourages and enables us to think differently, creatively and generate new ideas and approaches to solving problems.

Thought leadership works on knowledge and is also about learning. Learning is essential for creativity and innovation and thought leaders value learning. They also actively encourage the creation of a learning climate where creativity can flourish and innovation emerges.

Study buddy

Consider Ryde's (2007) repertoires with your study buddy. Compare and contrast your styles of leadership according to this model and share your findings with them. Help each other by identifying ways in which you can think more effectively.

3 Knowledge management and learning

Knowledge is a source of competitive advantage. In an era of information technology you might expect that we must be better than we have ever been at managing knowledge. Unfortunately many studies have shown the contrary is true.

A focus on technology has led to a tendency to focus on gathering data and because it is so easy, vast quantities of it can be gathered. A lack of selectivity in data collection can add to the problem but the main problem is that too often it remains as data never making the transition, through analysis and interpretation to knowledge.

Knowledge is vital for innovation whether we formally gather, analyse and disseminate knowledge or whether it is the informal process that goes on in our heads all the time. We use knowledge from a variety of sources to generate and build on ideas.

Hargadon and Sutton (2000) describe a 'Knowledge Brokering Cycle' that allows organisations to innovate time and time again, which might look something like this:

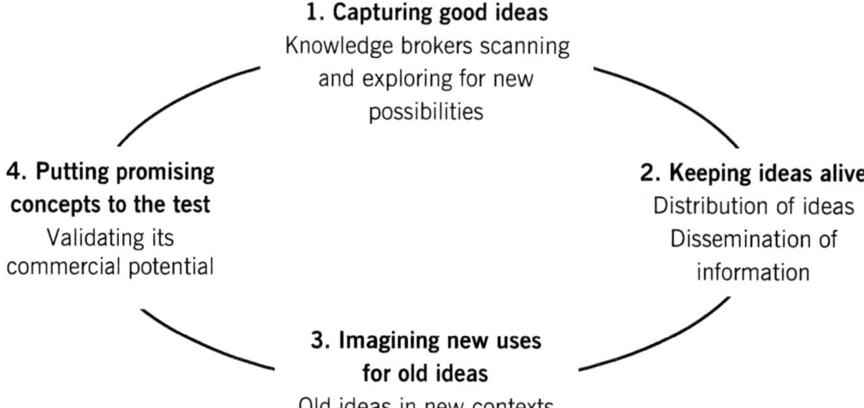

3.1 Knowledge ownership and creating the right climate

A good starting point is to establish the ownership of knowledge in an organisation, particularly in knowledge organisations.

There is no question that knowledge acquired and developed in the pursuit of organisation strategies and goals belongs to the organisation. However, the concept of shared ownership should create a culture of sharing and exchanging knowledge which is more likely to lead to innovation.

Policy and contractual arrangements will be needed to protect intellectual property but there should also be acknowledgement and recognition of those contributing to the knowledge pool.

If this is neglected a "knowledge is power" mentality can quickly emerge with people being protective of the knowledge they have and very selective about what they share with others. This is not conducive to innovation.

Creating the right climate requires the following:

Culture	Of freedom, constructive challenge and confrontation, as well as other factors already mentioned, are essential for knowledge sharing and innovation. The culture will also see people as central to knowledge and innovation and technology and process support activities.
Honest and open communications	Encourages the sharing and exchanging of knowledge. If employees are kept informed of decisions and events they are more likely to reciprocate with information.
People	It is not just about acquiring knowledge. People develop and use knowledge and this is where the real value exists. It is not simply a collection of disparate pieces of data but interpreted and stored as intelligence – connections are made and assumptions, conclusions drawn. The knowledge they hold is the sum total of this activity and the experience that goes with it. People are the experts in knowledge management. Staff should be motivated and rewarded for sharing and using knowledge.
Recognition and reward	Helps encourage the sharing and exchanging of knowledge and acknowledging the contribution employees make to the overall wealth of organisational knowledge. Performance measurement should take account of the value of knowledge and its contribution.
Processes	Provides the method for gathering, analysing, interpreting, disseminating and using knowledge. Good processes are kept simple and minimal If they are going to be effective. Processes should evolve to support knowledge management.
Technology	A tool to support knowledge management. An essential is the connectivity of technology across the organisation and its extension beyond to external stakeholders. The brief for technological requirements is not driven by finance and IT, it is a collaboration across the organisation resulting in a specification that meets marketing needs, collaboration with stakeholders and knowledge management and innovation needs.

3.2 A learning culture

Organisational learning goes beyond sending people off on training courses or internal management or other development programmes. Learning should be a way of life, a belief, a value integral to the organisation's culture.

Definition

Learning organisation: an organisation that actively develops and encourages the learning and development of all people in the organisation and integrates learning across the organisation with the goal of improving organisation performance and development.

3.3 Learning with a purpose

External training courses and learning and development programmes are all relevant. What sets apart good learning and development from poor is how connected it is to the organisation's goals, strategy and vision and how well the numerous learning events have objectives that have emerged from learning needs and whether these events are co-ordinated, integrated and evaluated for their outcomes.

Where learning and development is unintentionally designed in isolation of the workplace and no attention is paid to the transfer of learning, positive outcomes and benefits of the learning are either not realised or short lived. This can be demotivating for employees keen to make a positive difference and implement learning.

Attention to the transfer of learning is necessary, whether it is informal or formal learning, external courses or internal on the job learning.

Line managers should establish support systems, however simple, and together with the learner and colleagues, give thought to how the learning can be implemented into new work practices or as a change in behaviour. Work as usual is unlikely to result in the acceptance of learning.

Leaders can create a learning organisation starting with its culture. Leaders provide examples of learning behaviour and signal its importance through their commitment to objectives set and resources allocated.

A learning culture requires acceptance of mistakes, encouragement of exploration and investigation. Therefore cultures that are risk averse and strong on maintaining the *status quo* will struggle to develop a learning culture.

Learning comes from reviewing and evaluation of tasks and activities, for example on completion of a project a meeting to review and evaluate what can be learned from the experience can result in identifying improvements for the next project.

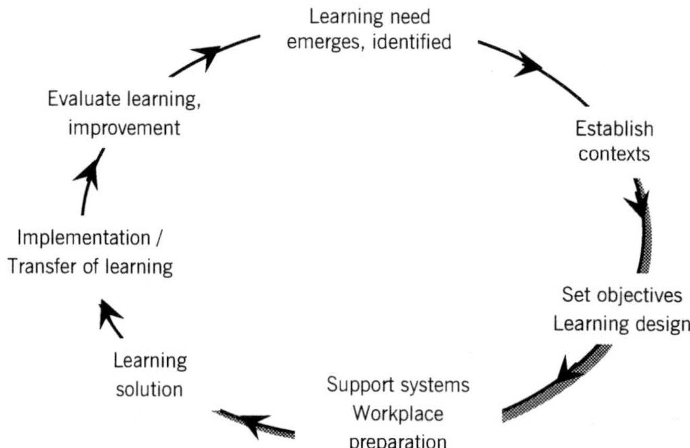

Systematic Learning and Transfer cycle

Activity 2

To what extent can your organisation be described as a learning organisation? Identify ways in which it is successful in this regard and ways in which it can improve.

Global case study

An organization with a strong learning culture can adapt more quickly than its competitors. Although much has been written about the idea of the learning organization, much of it has lacked specific prescriptions, leaving managers unsure how to implement it. The learning organization has also been held back because the concept was communicated primarily to CEOs and senior executives rather than at managers of departments and units, and because there has been a lack of standards and tools for assessment.

First, leadership alone is not enough. Modifying leadership behaviour is important, but it's also crucial to install formal learning processes and cultivate a supportive learning climate. Second, managers must acknowledge and be sensitive to differences among departmental processes and behaviours. Third,

Aligning the organisation to the vision

comparative performance is the critical scorecard, not absolute scores in isolation. Finally, learning is multidimensional and it's a mistake for companies to concentrate all their efforts to improve learning in a single area.

http://www.manyworlds.com/exploreco.aspx?coid=CO6170812434848

Accessed 27 March 2010

The important point to remember is that training and development must be designed to improve the participants and therefore the organisation's capabilities to realise its vision.

4 Leadership and collaboration

Leaders are required to design organisations to be able to exploit opportunities for creative, innovative and competitive partnerships.

These partnerships are very different in nature so a "one size fits all" approach will not work. We discussed earlier the fluid nature of leadership style and this is also true for organisation design.

As collaborations have increased and market conditions changed so organisations have responded with new business models emerging. Interestingly, as organisations have become flatter (or claim to be) they have become broader through alliances and collaboration which can be rather more difficult to manage than a hierarchical structure.

As well as creating a learning organisation, leaders need to resource connectivity through technology to be able to work with collaborative networks and virtual self-managed teams.

Developing strategic networks and collaboration needs clear objectives, strategy, stakeholder relationship management and control, however, the nature of collaboration means control may be rather challenging.

However, some level of control is essential as there are obviously risks associated with collaboration including:

- Sharing sensitive information or intellectual property and the risk of it ending up in the hands of the competition, or in the marketplace before it should, with damaging consequences

- Losing innovative ideas to others

- Exposing sources of sustainable competitive advantage that are then leaked to competitors

Collaborations should therefore be assessed for potential risks and we need to consider how collaboration impacts on competitive position.

Collaborative networks may well be competing with other collaborative networks so leaders have to work with each other across network on consistency and collaborative advantage that establishes a distinct position.

Collaborations between competitors can change competitive positions either deliberately or unintentionally. Where deliberate, the decisions are about what competitive position and how operations and marketing change as a result. The real challenge here is how much each organisation wants to change or can change.

The unintentional is often a lack of thinking through consequences. For example, when many airlines formed strategic alliances (allegedly for the benefit of customers) some of these alliance decisions were clearly not thought through from a competitive position perspective.

Consistency and continuity are important factors in deciding to form a collaborative partnership, particularly where competitive position may be compromised.

4.1 Multicultural leadership in collaboration

Even if an organisation does not consider itself international there are very few organisations that are not influenced by, or in contact with, international markets. This is particularly true in an organisation that has collaborative networks.

Collaborative networks are global with potentially anyone contributing to some process and, increasingly, that process is innovation. Working with teams globally even if they are virtual still means leaders need to be culturally aware and skilled at managing the differences if positive outcomes are to be realised.

There is much evidence that the culture convergence feared by many has not happened and most cultures remain intact.

However, the interconnectivity of cultures has increased and managing those aspects that influence the organisation is the challenge. Managing any form of operation that involves different cultures requires leaders to manage the diversity and ensure the best emerges from the cultural differences.

For organisations with overseas operations there may be more control and need for a consistent corporate identity but, even if what is being managed is a collaborative network of people or enterprises not part of the organisation, cultural sensitivity is essential.

There are three approaches to managing different cultures.

1   **Ignore the differences and carry on as usual**

    This risks the best from the cultures being lost and ignoring can result in cultural insensitivity leading to focusing negatively on the differences.

2   **Eliminate the differences and drive conformity**

    First, it is usually impossible to eliminate the differences and any attempts to do so will antagonise, demoralise and result in active or passive resistance or withdrawal.

    Second, while it might be desirable for good reasons to have a "one way of doing things around here" the focus should be on operations and harmonising what is commonly shared rather than stifle the diversity that can bring positive contributions.

3   **Accept the differences and work on commonalities**

    If groups are going to be working together for some time it is worth identifying commonalities early on, as this can avoid focusing on differences that may be perceived as threats, and move towards the differences focusing on the positive.

Processes and systems for doing things can be standardised but even here a consultation process together with negotiations and compromise should result in something everyone finds acceptable.

The emphasis for leaders is to encourage cultural co-operation by providing examples through their own behaviour. Their challenge is to manage diversity in a way that makes the most of all that is best.

For leaders to succeed cultural awareness is not enough. They need a deeper understanding of cultural differences and diversity and that means going beyond what is reasonably familiar to the edges of their organisation's world and what is unfamiliar. Being open and willing to engage with the unfamiliar, sends a positive signal.

5   Stakeholder relationship management

Keeping track of who the organisation is or should be collaborating with is not simple. Typically and most obviously an organisation is involved with:

- Customers
- Shareholders
- Financial institutions and other investors

- Distribution channels
- Suppliers

and in some cases:

- Trade association and industry institutions
- Regulatory and licensing bodies
- Analysts
- Non-government organisations (NGOs)

It is easier with competitor collaborations to identify who you might want to collaborate with and why. The challenge then is making it happen and making it work.

Identifying other collaborators in innovation may be more difficult. The possibilities are endless and then the challenge is one of being open enough to collaborate but balancing this with protecting intellectual property and creativity.

Stakeholders, other than employees, who collaborate in the innovation process include:

- Customers
- Suppliers and intermediaries
- Shareholders/members
- Trade unions
- Staff representative bodies
- Media and industry analysts
- Industry associations, regulatory bodies
- Universities, business schools

Mapping collaborative networks can be very useful for understanding who is involved.

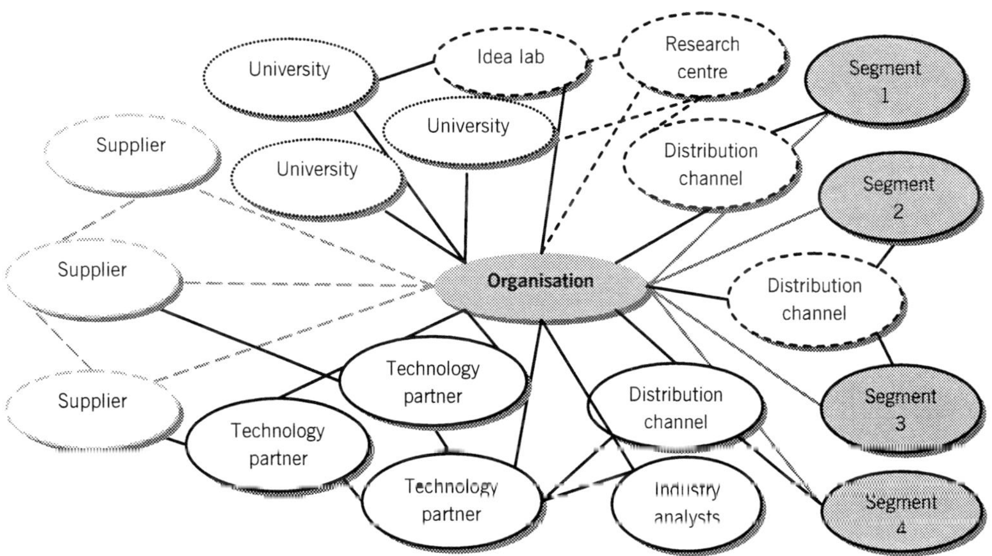

Collaborative innovation network or innovation communities

When all stakeholders of any collaboration have been identified the next step is to establish their power and impact.

5.1 Stakeholder influence and involvement

Stakeholders other than customers need to be understood and their impact on the organisation can be just as important. Some stakeholders' involvement with the organisation will be far greater than others as will their impact on value and will also vary from organisation or organisation. It can be useful to identify a stakeholder's level of involvement with the organisation and their influence on you.

Involvement with organisation
High Low

	High	Prioritise and actively manage	Actively manage
Direct influence on you			
		Monitor and manage relationship	Monitor

Factors that contribute to and define influence and impact include:

- Investment in organisation – (eg finance, knowledge or expertise)
- Authority over organisation – (eg government or regulatory body)
- Power of stakeholder – (eg pressure group or key distribution outlet)

Assignment advice

Think about stakeholder relationships to gain support for your assessment. You will require the support and co-operation of your managers and colleagues during assignments. Explain your needs and what is to be undertaken and any resources that may be required. It may mean returning the favour at some point!

5.2 What is involved in building and managing relationships?

Working relationships often fail because of a lack of strategy and management. They can fail or not work effectively due to misunderstandings, misinterpretations, conflicts and antagonism that can arise from poor communications and no effort to manage.

To work effectively there needs to be a formalised process for building and managing a working partnership. We therefore need a plan and guidelines on managing the partnership.

5.3 Compatibility

To what extent do people share values, beliefs and practices? If there are differences these will either need to be acknowledged and allowances made or changes managed. If allowances or changes are not enforced conflicts can quickly emerge distracting people from the purpose of the relationship.

5.4 Partnership

The leader signals the emphasis is on co-operation and managers work on:

- Building the relationship
- Commitment to quality improvement
- Regular face-to-face meetings
- Collaborative effort – understanding each other's responsibilities and commitments
- Exchanging information (ideas etc)
- Sharing mutual goals and some risks
- Flexible arrangements

5.5 Boundaries and learning curve

As people begin to work together there will be a period of learning about the relationship itself. Allow time for the network to learn about each other.

Setting clear boundaries establishes the role of the network in achieving strategic goals, defining expectations and commitment. This avoids misunderstandings. It is rather like an induction for new employees and can shorten and speed up the learning curve and build loyalty to the network.

- **Objectives** – the desired outcomes clarified.

- **Role and defining responsibilities** – who is responsible for what, eg providing funds, changing systems, supporting and encouraging practice.

- **Reporting and communication** – formal written reports and their purpose, eg to inform, update, meetings and briefings, informal reporting lines and structure.

- **Agreeing review procedures** – timing, methods and monitoring.

5.6 Communications

Implicit in the partnership is the need for good communications both formal and informal. As well as communicating expectations, feedback and exchanging information, the following, established at the start will provide a foundation.

- **Communications** – used to inform, educate, promote and persuade. For example informing people of the standards expected, promoting best practice.

- **Establishing the standards** – how they are identified, how they are validated.

- **Knowledge and information** – sharing and exchanging, using the best, selectivity and processes.

- **Leading by example** – being advocates of best practice and demonstrating through everyday practice.

5.7 Control and evaluation

Feedback should be part of the communications process. To what extent can existing internal controls be used? Do new controls need to be established? You would need:

- **Evaluation** – commitment to meeting standards by measuring the effectiveness of performance and establishing whether or not it has met the standard.

- **Standard of reporting** (ease of understanding, does it enlighten, address issues, are recommendations actionable?).

- **Impact on relationships** (steady and consistent or surprises and crisis).

- **Partnership development** (knowledge of each other's activities etc).

- **Added value** (what benefits have been derived from the partnership).

This provides us with a framework for forming and managing partnerships, that clarifies what people are expected to do. Without this formalisation stakeholders may be disinclined to take responsibility for outcomes.

Our approach to managing partnerships must be flexible and will form and re-form over time as different needs and change is implemented. We should also ensure we have mechanisms in place for learning from the partnership, what worked well, what was problematic and so on.

An end of project review of the actual partnership provides an opportunity to honestly and openly discuss how effective partnerships were with a view to improvement for next time.

How effectively partnerships work, particularly collaboration, will depend on the culture of the organisation. Some cultures will act as inhibitors to the effectiveness of the partnership. A driver of effective partnerships will be a learning culture.

Activity 3

Identify key stakeholders you collaborate with either informally or formally. How could you improve this collaboration and what would the benefits be?

6 Your leadership style

Now you have had time to think about and evaluate different leadership styles, motivation and other organisation factors that affect leadership you can start to work on your own career progression.

6.1 Developing your leadership style

Following the identification of your leadership style you can then set personal objectives and goals. In developing your leadership style there are a number of factors you need to take into account including:

- Of the leadership styles described what style/s would you like to develop, what sort of leader do you want to be?

- What is appropriate and possible in your organisation?

- What are the barriers to developing your preferred leadership style, internal and personal to you and within your organisation?

- What are the opportunities to develop leadership skills, for example leading projects, learning and development programmes?

Leadership style is strongly influenced by organisation culture.

Activity 4

Develop a Personal Development Plan (PDP) to improve an aspect of leadership you particularly want to improve. You do not have to be in a senior management role to do this. This is about improving leadership skills whether you are in already such a role or preparing for a leadership role.

7 Culture defined

Definition

Culture is the underlying, internal values, beliefs and attitudes that shape behaviour and the structure of perceptions. Culture goes deep and is stable. It is often difficult to assess and change culture.

7.1 Characteristics of culture

Culture is often described as "*How we do things around here*" (Deal and Kennedy, 1982) and the description certainly reflects the outward manifestation of culture. It can be deeply rooted in the history of

the organisation and its past achievements, trials and tribulations. Regardless of the country of origin, sports teams and governments visibly demonstrate this. Understanding culture is essential for leaders as not only can they have the greatest influence on culture but they can be influenced by culture and not always positively.

The destiny of the organisation is in the hands of leadership and one of the most enduring ways they can effect desirable behaviour and performance is through culture.

The make up of organisation culture is:

Artefacts and behaviours

- The most visible and superficial expressions of culture, for example, organisation practices, structures, systems rituals, routines, stories, myths, jokes, heroes and symbols, atmosphere.

- Artefacts are easy to observe but difficult to decipher.

Beliefs, values and attitudes

- These are part of the foundation organisational culture.

- Beliefs are what people think is true or not true, what is important or not important.

- Values are what people hold to be right or wrong, good or bad, their ethical code of conduct, for example, integrity, transparency and fairness.

- It can be difficult to recognise the difference between beliefs and values because beliefs about how to do something are often drive by values. For example, if people believe the only way to conduct business is by being honest, open and accountable is this based on their values or beliefs?

- Attitudes make the link between beliefs and values with feelings.

Underlying assumptions

- These at the deepest level and the most difficult to reach.

- Assumptions are mainly unconsciously formed by values over time as we interpret and make sense of our experiences and the world around us.

- This unconscious interpretation emerges as deeply embedded assumptions that guide perceptions, feelings and emotions that people share.

- Schein (1985) suggests that when all these three levels are aligned a homogeneous culture exists and that organisations with a homogeneous culture are more likely to be successful than those that are culturally out of alignment.

- There has been much research on organisation, particularly since the 1980s and some useful and enduring typologies have emerged that help us describe and understand what sort of organisational culture exists and therefore what works for and against us in achieving organisation goals.

Assessment advice

While planning your assignment it can be tempting to "sort the problem" and forget to incorporate the learning and/or try and shoehorn a loosely related work project into an assignment.

8 Organisational cultural types

8.1 Typologies of culture

Four typologies of organisation culture were developed by Charles Handy (1978).

8.1.1 Power culture

Power culture (or web) has few rules and regulations but there are strong policy decrees. Behaviour exhibited in such a culture includes aggressive, flexible and responsive.

A power culture is typically only as strong as the central figure who exercises control through a small circle of executives who usually retain financial control.

Politics are a feature of this culture and decisions are often based on political themes rather than for logical or operational reasons. Lack of formality means management development is difficult.

Personal development is achieved through a system of apprenticeship to the central power and this special place is dependent on the quality of the relationship. People are promoted because of who they know/get on with, not what they do/have achieved.

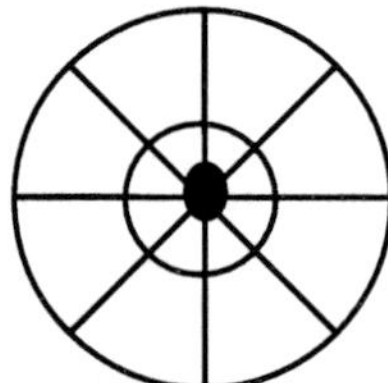

Perceptions of infallibility can lead to risk-taking and competition between employees is encouraged. A lack of rules, quick responses and competitive behaviour leads to a higher chance of surviving a crisis.

8.1.2 Person culture

A person culture is a cluster of individuals. The focus in this culture is on the individual and the organisation is not hierarchical, it only exists to support an individual or small number of individuals.

There is little structure imposed or control exercised in this culture. People usually come together to share administrative resources. Its lack of structure and formality is challenged as the business grows and, if it is to survive, the culture will eventually need to change to allow for growth.

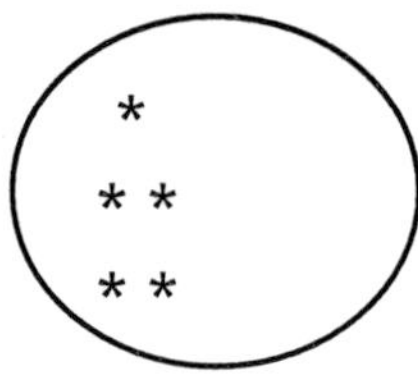

8.1.3 Task culture

Task culture (often described as a net) is a matrix structure which exists to support focus on the job. The organisation is task-oriented.

Power tends to be based on expert power. Teams form and re-form to complete specific projects and perform specific tasks. This ability to adapt and change makes the organisation very flexible and therefore suitable for rapidly changing market conditions. This culture tends to encourage innovation and creativity and people are judged by results.

Respect for people is based on their knowledge and expertise rather than status and position. There is high job satisfaction with an emphasis on group work. There is often no single individual in authority and power often lies in areas where the "net" is stronger and points cross.

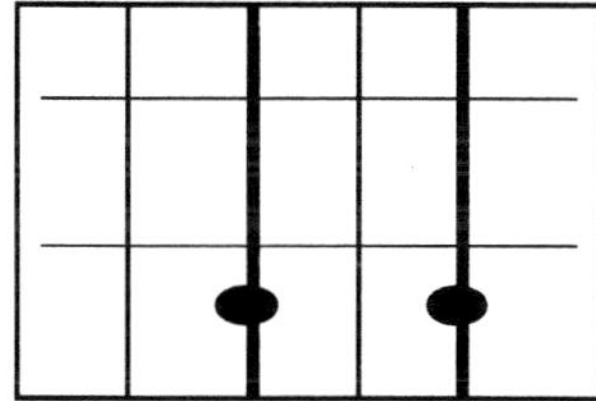

8.1.4 Role culture

Role culture is all about bureaucracy. It is risk adverse and rewards people for standard performance and not deviating from the norm.

A role culture can succeed in stable environmental conditions and where economies of scale are more important than flexibility. However, stability is rare, making flexibility essential, and economies of scale are easier to achieve with technological advances.

The focus on maintaining conformity usually results in the organisation losing the purpose for which it was formed. Systems, procedures, policies, rules and regulations have evolved over time and dominate the activities of this type of organisation. They help achieve highly valued predictability of performance.

There is a logical, rational arrangement to everything. Tasks are more important than people, which can dehumanise. Personal development is formalised.

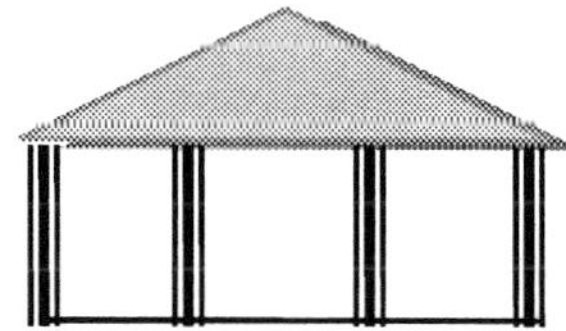

8.1.5 Cultures as competing values

Quinn and Rohrbaugh (1983) describe organisation cultures in terms of competing values that result in tensions and conflicts as follows:

- **The market**
  A rational culture where the boss is in control. Leadership style is directive and focused on goals achieved through productivity and efficiency.

- **The adhocracy**
  An idoological culture with charismatic leaders who are risk-takers. Leadership style encourages conformity to organisation values.

- **The clan**
  A consensual culture where relationships are important. Leadership style is supportive and encourages participation of employees.

- **The hierarchy**
  A hierarchical culture where leaders are risk-avoiders and control through formalised evaluation of performance. Leadership style is formal adherence to regulations and maintaining stability.

- **Competing values framework**
  The competing values framework assesses to what extent the organisation focuses internally on, for example, people or operations or externally, for example, on customers and competitors.

  It also assesses who makes decisions. At the lower-end management make decisions and at the upper-end employees have the power to make decisions.

Flexibility and discretion

Clan	Adhocracy
Hierarchy	Market

Internal focus and integration — External focus and differentiation

Stability and control

Source: Quinn and Rohrbaugh (1983)

8.2 Culture as rites and rituals

Corporate cultures include rites and rituals.

Deal and Kennedy (1982) identified four types of culture based on two factors in the marketplace: the degree of risk associated with organisation activities, and the speed at which organisations received feedback on success of decisions or strategies.

They described their four cultures as:

Tough guy, macho	Individualists who frequently take high risks, receive quick feedback. Examples cited include police, surgeons, construction, management consulting, entertainment industry.
Work hard/play hard	Fun, action, employees take few risks, quick feedback. High level of relatively low risk activity. Highly dynamic, primary value centres on customers and their needs. Teams produce volume, culture encourages games, meetings, promotions, conventions, to help maintain motivation. Volume can be at the expense of quality. Examples include sales organisations and mass consumer companies.
Bet your company	Large stake decisions, high risk but slow feedback. Focus on the future and importance of investing in it. Deliberateness is typified by the ritual of the business meeting. There is a hierarchical system of authority and decision-making is top-down. High quality inventions, breakthroughs, slow and vulnerable to short-term fluctuations. Examples: oil companies, investment banks, military.
Process	Low risk, slow feedback, employees find it difficult to measure what they do and get very little feedback on effectiveness so focus on how they do things not what they do. Individual financial stakes low. Memos, reports disappear into void. Cover your back mentality. Bureaucracy results in attention to trivia, detail, formality and technical perfection. Examples are banks, insurance, financial services, civil service, government.

 Aligning the organisation to the vision

8.3 Cultural web

Johnson and Scholes (2002) developed a framework for how cultures are created. This culture web, illustrated below, represents the "taken-for-granted" assumptions or paradigm of an organisation and the physical manifestations of organisation culture.

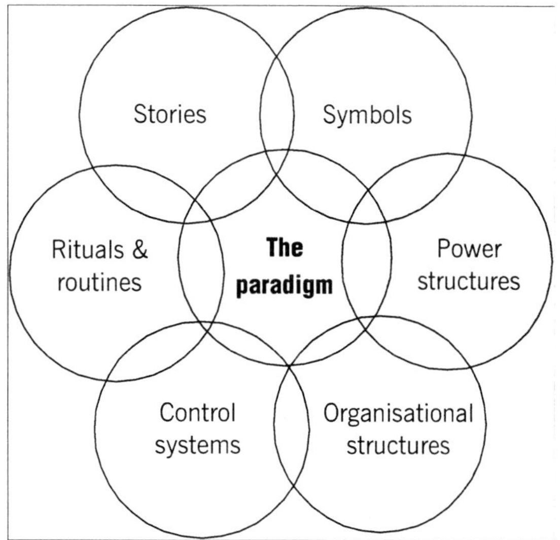

- **Stories**
  Link the past to the present, those immortalised, confer "status" on individuals (positive or negative).

- **Symbols**
  Physical evidence, eg logos, buildings, language and titles.

- **Power**
  Based on seniority and/or expertise, pockets of real power, most influence.

- **Rituals & routines**
  Special events that confirm and reinforce culture. Behaviour of members of the organisation towards each other, and external stakeholders.

- **Structure**
  Establishes relationships, reporting lines, unwritten lines of power, influence that indicate whose contributions valued, communication process.

- **Control**
  Design and focus of quality, reward systems and measurement

- **The paradigm**
  The sum of the other elements of cultural web, the concept and principles of the culture.

8.4 Sub and counter cultures

The existence of different groups within the one organisation can lead to disunity and conflict as there could be different cultures operating across the organisation. It may be that one or more groups have become disaffected and may develop objectives that run counter to those of the dominant group. They can create hostility and resistance to new ideas.

Evaluate the culture of your organisation. Can you identify aspects of cultures described and does one culture or different cultures exist? What are the positive and negative aspects of your culture?

9 Functional orientations and cultural drivers

The industry an organisation belongs to, the technology it uses or products and services it produces influence the organisation's culture.

Functional orientations are common. The following are some examples and while we are looking at extremes, which can be found in organisations, more typically an organisation will have one or more of these orientations that its reflects behaviour to some degree. The descriptions therefore provide some clues to what is influencing an organisation's behaviour.

These functional orientations are typically woven into the organisation cultures, as described above.

9.1 Product orientation

A technical- (product/service expertise) oriented organisation believes its products or services speak for themselves and therefore there is no need to worry about focusing on customers, they will inevitably beat a path to the door.

Initially, a gap in the market or a unique product sells itself in the early days so little or no effort is needed to market or sell the product providing customers are alerted in some way.

As markets become more competitive selling the product becomes more challenging. The response is to focus on the product and see how much better it can be made.

More quality, more reliability, for example, are all possibilities regardless of what customers want. The core values of a technical orientation might include:

- Technical ability/expert
- Innovative, quality and precision
- Idiosyncratic, quirky and eccentric

Some organisations see the development of ideas as their purpose, not customers' needs, wants and preferences. The prevailing values and attitudes affect their approach to a number of key issues.

Issues	Comments
People (management and staff)	Senior management will probably be technicians and there will be an emphasis on technical skills and expertise.
	Other skills, for example, finance, HR or marketing will be seen as either a necessary evil or unnecessary.
	They will find it difficult to trust those who do not understand technical expertise or engineering to run any aspect of the business.
Planning of activities	Allocation of resources will favour research and development and be internally focused on producing the product.
	Measurement focuses on product value and quality control checks.
	There will be an emphasis on efficiency and they will probably be good at sourcing quality materials.
	They can reassure themselves they are right because they have generated the ideas in the first place without help from customers or marketing.

Aligning the organisation to the vision

Issues	Comments
Communications	Will be "Features" led and one way, customers need to be informed on technical detail, imparting information rather than engaging customers or stakeholders through communication.
	There is no integrated communications and the product is the brand.
Customer service	Technical back-up and service will be good and "after-sales" service seen as an essential element.
	Not that their perfect product will need much attention, it is more about the technical back-up and expertise.

Among the key challenges and concerns for marketing leadership when changing the culture from a technical to a marketing/customer orientation are:

- The reduced power in new product development – marketing co-ordinates using information from customers.

- Quality goals maintenance – quality will be from the customers' perspective not an internally driven decision.

- Investment in research and development – will be competing with investment in marketing, research and communications, not previously seen as important.

Reassurance and removing perceived threats, improving customer knowledge as the driver of the creation and production of products and services should be the focus of managing the champions of technical orientation.

9.2 Sales orientation

A sales orientation believes that markets are highly competitive and the only way to succeed is to persuade, subtly or forcefully, customers to buy products and services.

Good sales people can sell anything so product quality is not necessarily a high priority. In some organisations acquiring new customers is given more prominence that retaining existing ones.

The core values of a sales orientation include:

- Winning, competitive, successful
- Aggressive
- Loyal
- Determined and persistent

It is difficult to succeed in this environment without these values. The prevailing values and attitudes affect the approach to a number of key issues.

Issues	Comments
People (management and staff)	Senior management will probably be sales people who have worked their way up through the organisation.
	There will be an emphasis on "charm" and personality and success comes from effective sales people.
Planning of activities	Allocation of resources will favour anything that supports the sales effort.
	Planning and measurement will be focused on short-term targets and volumes.
	Personal promises to customers, lack of long-term integrated planning and sales power means production targets will be shifted according to each sales person's "priority" and promise.
	Price is used as a tactical tool and has nothing to do with positioning.

Issues	Comments
Communications	There is no need for integrated marketing communications.
	Communications will be *ad hoc*, tactical and purely a support function for sales.
	There will be no plan or targeting as sales people are the main form of communications.
Customer service	Customer service is about personal friendships and can be very good for problem-solving and
	building relationships. Customer service will be sales activated and driven and therefore have a very personal
	interpretation. The result is *ad hoc* and tactical and the relationship is not built with the organisation.

Among the key challenges and concerns for marketing leadership when changing the culture from a sales to a marketing/customer orientation are:

- Reduced power in customer selection and targeting
- Reduced power in customer management and value creation
- Reduced power and influence in promotional activities

Customer management is shared and educating and persuading sales of the value of marketing and how it helps sales and the role of sales in marketing should be the focus of managing the champions of sales orientation.

9.3 Financial orientation

A financial orientation has become common as financial people rise to the top of organisations, often as a necessity because of poor financial management.

Once there, they do a good job of turning a financial crisis around. However, without either an innate understanding of customers or experience they continue to focus exclusively on the bottom line. The core values of a financial orientation may include:

- Certainty and prudence
- Meticulous and consistent
- Formulaic

The prevailing values and attitudes affect the approach to a number of key issues.

Issues	Comments
People (management and staff)	The CEO will probably be an accountant.
	They see people as an "overhead cost"
	and there is an emphasis on supervision and control.
	They are held accountable for mistakes so have a fear of losing control.
Planning of activities	Planning activities are actually budgeting periods and cycles and internal short-term focus is on the "bottom line".
	Cost plus pricing will be among the methods used to set price and position seen as irrelevant to price.
	Cost control is where their power base lies as is controlling budgets.
Communications	Communications will be one way and thought of as "if we must haves" such as brochures, a web site and so on.
	There will be little planning or targeting.

Issues	Comments
Customer service	Customer service is seen as luxury and a cost.
	Some aspects of service are seen as unnecessary (eg courtesy calls, personal mailers and so on).

Among the key challenges and concerns for marketing leadership when changing the culture from a financial to a marketing/customer orientation are:

- Reduced power in control with the introduction of a more comprehensive approach
- Long-term goals are not sacrificed for short-term gains
- Budget decisions reflect customer goals and objectives

Redefining and designing control and encouraging co-operation should be the focus of managing the champions of financial orientation.

9.4 Technological orientation

There is increasingly a technology orientation emerging (regardless of whether or not it is a technology organisation) where the solution to all problems is to buy more technology. This tends to be implemented without proper evaluation of the specification needed by all functions and the impact from a customer perspective. Thus the people who suffer most are customers.

No organisation culture or orientation is exactly as described above and, while we have explored the excesses of functional orientations, it is not unusual to find some symptoms of a particular orientation, which can provide managers with valuable insights into the implications on management style in the organisation.

Organisations, even small ones, do not conveniently have just one culture or orientation. Often a number of cultures can coexist and the larger the organisation the more likely it is that a number of cultures and orientations will exist.

When we understand our current culture and orientation we can assess the extent of change needed to achieve a marketing orientation. Marketing orientation is evolving and, increasingly, including a more societal orientation with a focus on corporate social responsibility, collaboration and partnership.

Discussion

Study buddy

Compare and contrast the orientation of your own organisation with your study buddy. Compare and contrast how each orientation supports and holds back a culture of innovation and entrepreneurship. Share your insights in the discussion forum.

10 Creating a marketing-oriented culture

A marketing-oriented culture must, by design be innovative and entrepreneurial. If we think about what marketing should be doing for the organisation and its customers it helps us focus on what a marketing-oriented culture should be.

Issues	Comments
Intelligence and knowledge	Through marketing research and knowledge management marketing can ensure the organisation is aware of macro changes and possible impacts, aware of current and future trends, patterns and possibly scenarios, knowledgeable about and understand customers, alert to competitor behaviour and possible consequences, informed in its decision-making and cognisant of the consequences of various decisions and actions.

Issues	Comments
Value creation	From this knowledge marketing can guide the organisation in its strategy development and value creation avoiding or reducing costly new product/service failures and increasing the potential success of new product/service launches. It can create points of differentiation that match organisation design and customer needs and influence changes in organisation design if needs are not met or are changing.
Communication	Through marketing communication marketing can strategically create, build and maintain corporate reputation and the brand. It can support competitive positioning, operationally create, build and maintain product/service range brands. Relationships with key stakeholders and are built while tactically creating and stimulating demand for products and services. Marketing can also reinforce external messages with internal communications that aligns behaviour with desired corporate and brand values, informs staff of external market expectations and their role in meeting those expectations.

10.1 People, integrity, trust and respect

The discussion on organisation culture covered many examples of the sort of values an organisation and its people might hold and how this affects the way people behave towards and treat each other.

If a marketing orientation is going to be achieved a review of values is a good starting point. An organisation might make statements about its desired (or actual) values but if the behaviour that customers experience is poor and incompetent and does not match the stated values customers will not, and should not, believe in those stated values.

Customers expect, and are increasingly demanding, that organisations who want customer loyalty behave with integrity, ethically and, at the very least, treat them with respect.

These are inescapable facts if an organisation genuinely wants to be perceived favourably by customers, employees and other key stakeholders. Marketing must be taken seriously and represented at board level either by a CEO who understands and practices a marketing philosophy and/or by qualified marketing professionals.

People are valued and trusted by managers and this is reflected in internal policies and procedures and management attitudes and behaviour to employees. It is recognised that people are key to success and a source of advantage.

The values of the organisation are reflected in how responsive the organisation is to the local community and the responsibility assumed for social and environmental concerns. Customers are the responsibility of everyone in the organisation. Each employee understands their role and contribution in delivering benefits to customers and knows who their customers are and what is important to them.

10.2 Marketing information

Knowledge is recognised as highly valuable and a source of advantage in a marketing-oriented organisation. Technology does not dictate marketing intelligence, knowledge management processes or design, it is a tool.

Leaders, managers and employees discuss, negotiate and agree what technology is required and how it is used and the focus is on converting gathered data into useful intelligence. Technology supports the use of knowledge in building relationships with customers and stakeholders.

10.3 Value creation, creativity and innovation

Creativity and innovation are encouraged and valued. The development of new products or services is undertaken with customer input either directly or through marketing intelligence. However, marketing also collaborate with others in the organisation to develop ideas.

Organisation culture can stifle or suppress creativity. We often hear of organisations that have a "blame" culture and the very negative impact this has on creativity and taking initiative. A blame culture is typified by accusations, threats, criticism and destructive feedback. An approval culture encourages experimentation and is forgiving of mistakes. It is typified by consent, praise and constructive feedback.

Leaders and managers in a marketing orientation create a climate for creativity through freedom, time, space and a willingness to allow people to explore, challenge the *status quo*, to think differently and be different. However, this creative climate is not designed to encourage creativity for its own sake, it has purpose and usually the objective of a clear commercial outcome.

The difference between creativity and innovation is understood and for innovation there is a process for evaluating and testing ideas for their commercial viability. The way this works is for different people with different characteristics and skills to form teams for different parts of the creativity and innovation process.

Quality and customer service are seen as integral to product and service development, not an after thought or add on. The clear objective is to create and maintain competitive advantage in order to acquire and retain customers through enhanced value. This value is represented in relevant and meaningful quality, customer service and brand values which customers can relate to.

10.4 Strategies and planning

Senior management focuses on the future and long-term direction of the organisation.

In a marketing orientation strategies are developed around customers and reflect the reality of the marketplace. Marketing plays an influential role in determining what product/service market opportunities the organisation might pursue. Financial resource allocation for marketing is determined by the objectives to be achieved not spare cash left over.

Planning is ongoing, methodical but flexible and realistic and the long-term goals are not sacrificed for short gain. Planning and control is recognised as the most efficient and effective way to use limited resources and to organise activities.

A feature of marketing planning is the co-ordination and integration of organisation activities to deliver value to customers and stakeholders. Co-ordination ensures consistency of delivery and takes advantage of maximising value added opportunities. Planning is designed around customer needs so segmentation, positioning and targeting is important.

The marketing mix is integrated, co-ordinated and designed to meet needs.

Departments understand each other's problems and needs and work in partnership to solve these and accomplish overall departmental and corporate objectives. Suppliers, distribution channels and other stakeholders are included in the planning process and their activities assimilated into the organisation's plans.

Changing the measures is an important step in supporting a change to a marketing orientation. The most important point to make is that employees have customer satisfaction performance targets and leaders and managers are also accountable to customers and key stakeholders and for more than just the financial goals.

10.5 Learning

Learning from experience is important and informs future decisions and actions. Informal and formal learning are encouraged.

Training and development of employees is recognised as key to business success and marketing and customer service training is not confined to marketing personnel only. Training and development is also seen as the most effective way of ensuring employees are prepared and able to accept and implement change in response to market conditions.

Appraisals form part of continual improvement and development and motivation of employees viewed as a significant management task.

Global case study

A review by the UK-based Chartered Institute of Personnel and Development (April 2009) reveals that leaders' responses to the recession fell back on old internally focused methods of reacting to a crisis and in doing so are putting the long-term future of the organisation at risk. Leaders admit regretting the cost-cutting measures taken during the last recession but are reacting the same way this time. This will lead, as last time, to skills shortages, low employee morale and customer dissatisfaction and attrition. Executives were asked which cost-cutting measures worked during the last recession and admitted that cutting back on training, incentives and marketing events had hampered employee commitment and seen their markets shrinking and not necessarily because of the recession.

Employers who hired high performing employees from competitors had the most positive impact on employee engagement and organisation performance during the last recession but is one of the least popular actions.

10.6 Communications

Communications both internally and externally with employees, customers and all stakeholders incorporate values and positioning and reflect the needs of different target groups. Two-way communications are important and there are mechanisms for listening as well as talking.

Marketing communications make promises and a marketing orientation ensures the promises are kept. Integration of marketing communications across internal and external markets ensures promises and the brand position is based on genuine values and behaviour.

10.7 The role of the 'intrapreneur'

Intrapreneurship is a strategy for stimulating innovation by making better use of entrepreneurial talent. When effectively promoted and channelled, intrapreneurship not only fosters innovation, it also helps employees with good ideas to better channel the resources of a corporation to develop more successful products.

Some of the greatest business leaders of the past century made their early mark in business as intrapreneurs. Former General Electric chairman Jack Welch made a name for himself by building GE's engineering plastics business as if he were starting his own company. Lew Lehr, former chairman of 3M, similarly built his career on his intrapreneurial pursuit of 3M's expansion into the healthcare industry.

By fostering an intrapreneur ethic within a company, employees can be empowered and enabled to become company 'change agents' who are comfortable bringing new ideas forward and promoting their execution.

It is essential to create an elevating and encouraging environment that provides talented and entrepreneurially-minded people with the freedom to innovate, while at the same time supporting them with the resources to quickly bring their innovations to market. Creating, fostering and sustaining the right environment really are intrapreneurial imperatives.

A key element of intrapreneurship is the ability of a company to support, with economic and technical resources, expedited decision-making processes. Furthermore, it should be able to demonstrate the willingness to break with traditions by embracing initiatives that run counter to the way the company has done things in the past.

Tom Nies, CEO of Cincom is the longest active-serving CEO in the computer industry and recognised alongside Bill Gates and Steve Jobs as one of the 'pioneers of the software industry'. An example of how this has worked in practice in his own business, Cincom Systems is the development of their call centre offering. Despite a long background in working for software developers and large telecommunications firms, one of his staff brokered the marriage between Cincom's technology and customer base and the growing need to reduce the cost of customer service to create an outsource call centre business. He gave this vision his personal support by travelling to India to promote the opening of their call centre business there. In such ways, companies can continue to find new means to leverage resources to create new business opportunities.

http://www.the-chiefexecutive.com/features/feature254/

Accessed 27 March 2010

Intrapreneurs thrive on the freedom which fuels their innate desire to innovate. This can be problematic for a manager who doesn't understand or respect the entrepreneurial nature. For intrapreneurship to flourish in an organisation, leadership has to be willing to listen to and recognise good ideas whenever and from whomever they arise. This message must be constantly reinforced from the highest levels of the organisation.

Beyond listening, it is critical to enable people to see their own ideas through, even if they must cross over into a new functional area and push themselves past any previous company achievements or organisational structure.

It is important to create an environment where anyone can come forward with an idea on how to improve any aspect of the business. It should not matter where that person fits on the organisational chart. If the idea is good, and the benefits and risks are clearly stated, that idea should get the green light - and the support it merits.

There must still be a business approval process, but it should be efficient. Projects that deserve support should be quickly expedited. So, for example, if it is not necessary to wade through a lengthy buy-in process, then it may be better to avoid this than to miss a window of opportunity.

Companies can foster and encourage potential intrapreneurs by sending the message throughout the organisation that a case properly presented, which thinks through the issues, identifies and explains what can go wrong, what can go right, and how to put contingencies in place, is welcome. But, the process must be simple and flexible enough to move quickly – and then to later scale up rapidly when success develops.

An intrapreneurial culture must embrace constructive failure to score big victories. The key is creating an environment where an employee's ideas, when properly presented, are taken seriously and then be properly supported and recognised. One never knows where good ideas will come from, especially in a corporate culture that supports intrapreneurship.

Many entrepreneurial careers are built on a succession of minor failures, with the accumulated lessons learned from each leading to ultimate success. It is important for companies to allow for a degree of inevitable failure around new projects and initiatives without sending the message that failure is not tolerated.

Companies must strive to provide a 'freedom to fail' culture and environment. Although failure resulting from poor planning and execution is not acceptable, there should be no penalty for those who come forward with good ideas, assuming they've been well presented and competently executed.

As stated earlier, an intrapreneurial culture must embrace constructive failure to score big victories. Many companies are filled with reliable 'singles hitters' who play it safe and never really aspire to greatness. Intrapreneurs, on the other hand, swing for the fences. Sometimes they strike out, but when they connect they like to hit it big.

It doesn't do any good to encourage team members at all levels to bring innovative ideas to company leadership if the leaders then take those ideas and make them their own. Recognition is a key driver for us all. Leaders who seek to steal the recognition rightfully deserved by others find few followers.

So, one needs to make sure credit goes where it is due, and is shared widely. It costs nothing to admit that the $10 million idea came from the receptionist. No one is diminished as a result, and the company is $10 million richer for it. The receptionist becomes then even more eager to offer better future ideas and, everyone else in the organisation is encouraged to follow the lead of that receptionist, and to help to improve the organisation.

As Tom Nies, CEO of Cincom says:

"Every organisation must have processes and rules of procedure and behaviour. But when we catch ourselves saying, 'we've never done it that way before," or 'that's not how we do things', we should stop and reflect on whether we are saying this out of habit, or for good reason. Chances are we may be citing a rule that may no longer be appropriate for the new conditions and situations we are now trying to intrapreneurially develop. Maybe it's best, and even necessary, to sometimes break with past traditions and establish new precedents to respond to new opportunities."

The ability to differentiate between rules needed to guide and perform within the current business and rules which may restrict success in building a new business is what discernment and opportunity awareness are all about. Going forward is always a journey, and as journeys progress we need new signposts along the way which point the way forward on the next leg of our trip. These signposts are the rules and regulations for building new businesses within existing businesses.

In light of the above critically evaluate how intrapreneurial your organisation is, what it is doing well and what it could do better. Try and identify at least one example of intrapreneurship that has contributed towards a business improvement

11 Shared values

As we have said values are central to any organisation and having the right values that everyone shares can be the difference between success and failure.

How to encourage everyone to share values is not particularly easy. Values, rather than beliefs and underlying assumptions, are what tend to be discussed most in organisations today and it has become increasingly popular to make declarations about what corporate values the organisation holds.

11.1 Corporate values and brand values

Stated values will always be positive but often some are aspirations rather than actual values that exist in the organisation.

It is an interesting exercise to take an organisation you know and see if the experiences you have of that organisation matches its stated corporate values. What often is not acknowledged or dealt with are the negative values that inevitably exist in many organisations. It is not enough to just make a statement on corporate values.

Positive stated values that exist must be protected and re-enforced and those that are aspirations must be explained, interpreted and encouraged through various examples, behaviours and programmes.

Internal marketing communication is a key means of encouraging leaders, managers and employees to share and live the values.

Recruiting like-minded people might be a good start and certainly has some merit. However there are advantages and disadvantages.

Advantages	Disadvantages
People sharing common values and, quite likely, similar beliefs and underlying assumptions makes the job of achieving a homogeneous culture much easier.	Everyone being of a similar mind can lead to a narrow perspective, and possibly a lack of creativity and innovation, such as an unwillingness to challenge when it is important to do so or to deliberately take an opposing view to stimulate a discussion or debate issues.

However, all sorts of factors can infect culture, particularly outside forces that might threaten the organisation so some proactive strategy to maintain the culture is still required.

11.2 Encouraging people to share values

A better starting point than stating values is to first establish what the real values are. This is an activity that organisations are increasingly engaging in. It also requires surveying internal employees, customers and other key stakeholders. All have different interactions with the organisation so will have valuable different perspectives. From this the organisation can establish commonalities, differences, positive and negative values.

Leaders may have clear ideas on what values they want the organisation to have but it is more productive to consult with employees and others to find out what they want. Apart from getting a better perspective on what is possible, it starts the process of engaging employees, in particular, in the task of getting them to share values.

From this exercise it is then possible to announce the stated values, being honest about what currently exists that is positive and to be protected, what is negative and needs attention and what is new to be adopted.

11.3 Making the link between corporate values, brand values and the customer

Marketers, in particular, have a responsibility for ensuring that there is a link between the customer, the value proposition, the brand values and organisation values.

Employees understanding what corporate values mean in their job do not automatically register statements issued from on high. Marketing managers are responsible for interpreting and translating corporate values into brand values and brand values into job activities, what it means and what is acceptable and unacceptable behaviour. Managers need to describe and illustrate what values mean for employees to be able to understand and act on delivering them.

Making the connection between brand values and what employees do is essential in matching customer experiences with expectations.

If one of our brand values is that we are "responsible" then employees must be able to take responsibility and expect to be accountable. Examples of what this means are demonstrated by leaders and managers

being held accountable for actions and outcomes. If a value states we are "reliable" then employees must always deliver on promises. Again, examples of what this means can be found in leaders and managers keeping their promises.

Some employees will get this quicker than others and will be leading examples of the brand values. Marketing leaders should identify and "recruit" these people to help reinforce and embed desirable behaviour and practice.

If we want employees to live the brand we must engage them and we cannot do that if they do not understand or believe in the values.

Leaders must:

- Explain the meaning of the brand – values, reflection, imagery
- Make an emotional connection – commitment
- Engender an enthusiasm to deliver – eg creativity, service, problem-solving
- Ensure people are competent to deliver – training and development

Ideally, leaders are examples of what the values mean and that people can emulate. However, if stated corporate brand values are in stark contrast to what is experienced daily by employees, the values will quickly lose meaning and conviction and employees will lose interest.

If leaders are not actively committed and never get beyond lip service marketers can attempt to influence a change. They can draw attention to examples of excellent leadership on corporate and brand values and how this has lead to strong reputations and customer loyalty. They can also point out the consequences of failing to commit to corporate and brand values. Marketers must become examples of good leadership in creating, building and maintaining values and behaviour that support brand values.

Subtlety and patience is important and timing and opportunity part of a deliberate campaign to change hearts and minds.

As others see repeated and consistent behaviour and leadership that is reflective of values and that this leads to positive real outcomes that match stated values, the gap between lip service and commitment becomes apparent and difficult to ignore.

Global case study

When, in 2007, the merger of Canada's Thomson Corporate, one of the largest international information companies and Reuters, the leading news and market data provider took place the leadership of the newly formed Thomson Reuters saw this as an opportunity to transform the marketing organisation rather than integrate two legacy marketing departments. The leaders started by defining and clarifying the goal for its new marketing organisation – to provide strategic vision and leadership for the Thomson Reuters business. This signals a very clear message to everyone in the organisation of the role of marketing.

The leadership was not leaving anything to chance. A thorough research project internally and externally preceded decisions on detail and revealed the challenges of achieving their goal. Among the challenges were the need to align resources with business priorities, build a common view of marketing's role, raise the profile and visibility of marketing, adopt common planning practices, clarify required competence and create progression plans for marketing staff. The re-orientation was seen as significant and while it was only given eight months it was a full-time project for a small leadership team. To secure marketing's strategic role a newly established Chief Marketing Officer was appointed and a "seat at the top table". Among the changes were changes to measures of performance reflecting both customer and strategic marketing performance.

 Aligning the organisation to the vision

1 Evaluate the role of leadership in encouraging innovation

- Organisation success and survival are dependent on innovation and creativity but this can often be stifled.

- Organisations need creativity that leads to innovation on new ways of doing things and new things to do. This is how businesses grow and prosper.

- Monitoring the external environment is part of the innovation process. Creativity responds to changing market dynamics.

- No organisation is exempt from the need to innovate. What differs is the speed of response and the nature and extent of innovation required.

- Leaders need to enthuse managers and employees with the task of finding both short-term innovation, often the marketing mix, and longer-term innovation, for example corporate strategy.

- No matter how successful or even if the organisation is a market leader leaders can never be complacent. A leader should never lose sight of the need to nurture creativity.

2 Consider the role of thought leadership and knowledge management in organisations

- Thought leadership is shared by all who can think creatively or differently.

- Thought leadership works on knowledge and is also about learning.

- Learning is essential for creativity and innovation and thought leaders value learning.

- Thought leaders actively encourage the creation of a learning climate where creativity can flourish and innovation emerges.

- Knowledge is a source of competitive advantage. It is vital for innovation whether we formally gather, analyse and disseminate knowledge or whether it is the informal process that goes on in our heads all the time.

- Creating the right climate requires a number of features including honest and open communication, supportive culture, reward and recognition systems, technology and so on.

- Organisational learning goes beyond sending people off on training courses or internal management or other development programmes.

- A learning organisation actively develops and encourages the learning and development of all people in the organisation and integrates learning across the organisation with the goal of improving performance and development.

- Leaders can create a learning organisation starting with its culture.

3 Identify and explain why collaboration is important and how to manage it effectively

- Leaders are required to design organisations to be able to exploit opportunities for creative, innovative and competitive partnerships.

- As collaborations increase and market conditions change so organisations have responded with new business models emerging.

- Developing strategic networks and collaboration needs clear objectives, strategy, stakeholder relationship management and control. The nature of collaboration means control might be rather challenging.

- Consistency and continuity are important factors in deciding to form a collaborative partnership, particularly where competitive position may be compromised.

- There is much evidence that the culture convergence feared by many has not happened and most cultures remain intact. For organisations with overseas operations cultural sensitivity is essential.

- The emphasis for leaders is to encourage cultural co-operation and manage diversity in a way that makes the most of all that is best.

4 Explore how stakeholder relationships can be managed

- Keeping track of who the organisation is or should be collaborating with is not simple.

- Identifying collaborators in innovation may be difficult. The possibilities are endless and then the challenge is one of being open enough to collaborate but balancing this with protecting intellectual property and creativity.

- Mapping collaborative networks can be very useful for understanding who is involved.

- When all stakeholders of any collaboration have been identified the next step is to establish their power and impact.

- To work effectively there needs to be a formalised process for building and managing a working partnership.

- Setting clear boundaries establishes the role of the network in achieving strategic goals, defining expectations and commitment.

- Implicit in the partnership is the need for good communications both formal and informal.

- Feedback should be part of the communications process.

5 Identify your desired leadership style

- Once you identify your leadership style you can set personal objectives and goals.

- Leadership style is strongly influenced by organisation culture.

6 Define organisational culture

- Culture is often described as "How we do things around here" (Deal and Kennedy, 1982) and the description certainly reflects the outward manifestation of culture.

- One of the most enduring ways leaders can effect desirable behaviour and performance is through culture.

- The make up of organisation culture is artefacts and behaviours, beliefs, values and attitudes and underlying assumptions.

7 Evaluate cultural frameworks and models

- Four typologies of organisation culture were developed by Charles Handy in 1978 and still hold true today: power, person, task and role culture.

- Quinn and Rohrbough (1983) described organisation cultures in terms of competing values that result in tensions and conflicts.

- Corporate cultures include rites and rituals.

- Johnson and Scholes (2002) developed a framework for how cultures are created called the 'cultural web'.

- The existence of different groups within the one organisation can lead to disunity and conflict. They can create hostility and resistance to new ideas.

8 Compare organisational orientations

- The industry an organisation belongs to, the technology it uses or products and services it produces influence the organisation's culture.

- Functional orientations are common. Most organisations fall into one of four categories: product orientation, sales orientation, financial orientation and technological orientation.

- No organisation culture or orientation is exactly the same but can provide managers with valuable insights into the implications on management style in the organisation.

- Organisations, even small ones, do not conveniently have just one culture or orientation.

- When we understand our current culture and orientation we can assess the extent of change needed to achieve a marketing orientation.

- A marketing-oriented culture must, by design, be innovative and entrepreneurial.

- If a marketing orientation is to be achieved a review of values is a good starting point. Customers expect, and are increasingly demanding, that organisations behave with integrity, ethically and at the very least treat them with respect.

- People are valued and trusted by managers and this is reflected in internal policies and procedures, management attitudes and behaviour to employees.

- The values of the organisation are reflected in how responsive the organisation is to the local community and the responsibility assumed for social and environmental concerns.

- Knowledge is recognised as highly valuable and a source of advantage in a marketing-oriented organisation. Technology supports the use of knowledge in building relationships with customers and stakeholders.

- Organisation culture can stifle or suppress creativity. In a marketing orientation strategies are developed around customers and reflect the reality of the marketplace.

9 Critically evaluate the importance of shared values

- Encouraging everyone to share values is not particularly easy. It has become increasingly popular to make declarations about what corporate values the organisation holds.

- Positive stated values that exist must be protected and re-enforced.

- Aspirations must be explained, interpreted and encouraged through various examples, behaviours and programmes.

- Leaders may have clear ideas on what values they want the organisation to have but it is more productive to consult with employees and others to find out what they want.

- Marketers have a responsibility for ensuring that there is a link between the customer, the value proposition, the brand values and organisation values.

- Making the connection between brand values and what employees do is essential in matching customer experiences with expectations.

- If leaders are not actively committed and never get beyond lip service marketers can attempt to influence a change. Subtlety and patience will be important and timing and opportunity part of a deliberate campaign to change hearts and minds.

Activity 1

You may identify a mix of the external and internal barriers we discussed in this chapter. You may also identify triggers of change that seem to have come from no where, do not seem to have been in response to an obvious external or internal trigger. If this is the case what was the trigger? Was it the result of a leadership whim, an internally created problem? If you identify internal barriers, what is the nature of these barriers, do these barriers keep recurring, is it possible to eliminate or reduce these barriers and to what extent have efforts been made to do this? Finally, what was the impact, did it force people to look for creative ways of managing or dealing with the barrier, was it ignored, did it actually prevent change or damage its success?

Activity 2

A learning organisation connects development opportunities to the organisation's goals, strategy and vision. Leaders create a learning organisation starting with its culture. Leaders provide examples of learning behaviour and signal its importance through their commitment to objectives set and resources allocated. A learning culture requires acceptance of mistakes, encouragement of exploration and investigation.

Activity 3

For many organisations collaboration might be informal and this may be appropriate. Where it is not appropriate is where the organisation's core competence or competitive advantage is dependent on the collaboration. The task is to identify the objectives of the collaboration, the desired outcomes and then establish measures to ensure those outcomes are realised. Building good relationships with those with whom the organisation collaborates is essential and communications in various forms the best way of achieving this. Benefits will very much depend on your own particular situation but may include for example, increased customer added value, improved core competence or innovation, reduced risk and costs.

Activity 4

In developing your PDP you might include improving conceptual skills, for example and working on creatively interpreting the future and a vision for your organisation for that future. Your improvements could work on improving your ability to motivate your staff or even work with peers to improve overall motivation.

Activity 5

There are many models and frameworks that are generally accepted as descriptions and characteristics of cultural type. The answer to this question will depend to a large extent on your organisation. Using accepted frameworks will help you critically evaluate the strengths and weaknesses as well as the opportunities and threats to creating an organisation aligned to the vision.

Activity 6

This answer will depend to a large extent on your organisation and the degree to which it encourages 'intrapreneurship'. With reference to the key features identified you could evaluate on a scale 1–5 how closely your organisation has the features required and then recommend strategies where there is the greatest need for improvement.

Amabile, T. M., (1998). 'How to kill creativity'. *Harvard Business Review,* September–October.

Cockton, J., (2002). *Systematic Learning and Transfer Cycle*. London: CIPD.

Deal, T. E. and Kennedy, A. A., (1982). *Corporate Cultures: The Rites and Rituals of Corporate Life*. London: Penguin.

De Bono, E., (1970). *Lateral Thinking*. London: Penguin.

Handy, C., (1978). *Gods of Management: The Changing Work of Organizations*. Oxford: Oxford University Press.

Hargadon, A. and Sutton, R. I., (2000). 'Building an innovation factory'. *Harvard Business Review*, May–June: p 57–166.

Johnson, G. and Scholes, K., (2002). *Exploring Corporate Strategy*. Oxford: Prentice Hall.

Ryde, R., (2007). *Thought Leadership*. London: Palgrave.

Schein, E., (1985). *Organisational Culture and Leadership*. San Francisco: Jossey Bass.

Quinn, T. and Rohrbough, D., (1983) 'A spatial model of effectiveness criteria: towards a competing values approach to organisational analysis'. *Management Science,* Vol. 29, p.363–77.

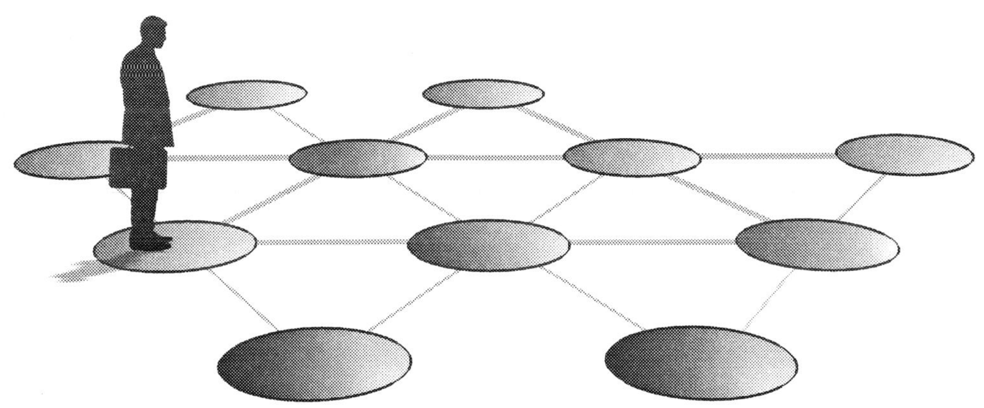

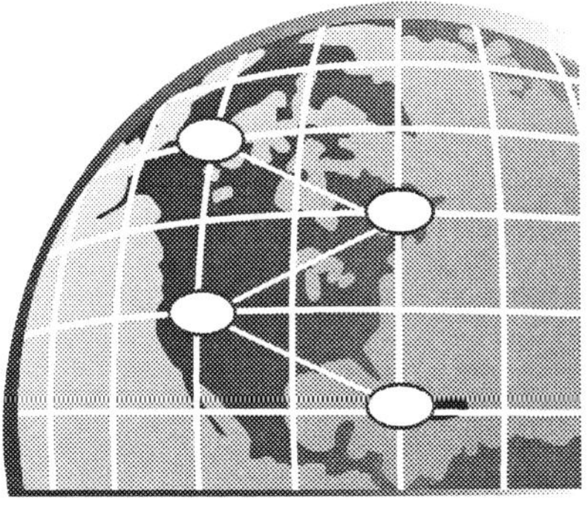

4 | chapter

Organisational structure and the role of quality management in aligning the organisation to the vision

This chapter looks at two key themes in aligning the organisation to the vision: quality management and structure.

We start by outlining the functions and elements of formal organisational structure and its impact on division of work, authority, relationships and job roles. We consider the various influences that shape organisational structure over time. Structures can be intentionally designed to support the delivery of marketing value, focus and creativity.

Quality approaches when integrated within a well functioning structure help an organisation achieve its vision and given that successful visions need to be marketing-oriented and must seek to satisfy customers with high performance. Service marketers can learn a great deal from the discipline because a quality approach instils a sense of responsibility in us as managers and leaders and across the organisation.

In our discussions we outline some key quality concepts, including various definitions of quality, and the dimensions and aspects of quality. We put forward the business case for quality: the reasons why it is important for organisations to build and maintain quality, and systems for pursuing and controlling quality. We also introduce some of the main quality strategies which an organisation might adopt when aligning itself to the vision.

Contents

Chapter learning outcomes

In this chapter you will cover the following:

- Evaluate the importance of organisational structures in delivering marketing value, focus and creativity that aligns the organisation to the vision

- Recommend how an organisation could be structured to deliver competitive advantage and organisational success

- Evaluate the issues that need to be taken into consideration in designing an organisation structure that aligns itself to the vision

- Assess the nature of quality

- Evaluate quality systems and processes

- Understand the concept of Total Quality Management

- Explore quality improvement approaches

- Critically evaluate the role of service quality

1 An overview of organisational structure

> **Definition**
>
> **Organisation structure** may be defined as 'the pattern of relationships among positions in the organisation and among members of the organisation. Structure makes possible the application of the process of management and creates a framework of order and command through which the activities of the organisation can be planned, organised, directed and controlled.' (Mullins, 1999)
>
> Mintzberg (1984) defines an organisation's structure as: 'The sum total of the ways in which it divides its labour into distinct tasks and then achieves co-ordination among them'.

Organisation structure thus implies a framework intended to:

- Define work roles and relationships, so that areas and flows of authority and responsibility are clearly established.

- Define work tasks and responsibilities, grouping and allocating them to suitable individuals and groups.

- Channel information flows (communication) efficiently through the organisation.

- Co-ordinate the objectives and activities of different units, so that overall aims are achieved without gaps or overlaps in the flow of work.

- Control the flow of work, information and resources, through the organisation.

- Support flexibility and adaptability to changing internal and external demands.

- Support the commitment, involvement and satisfaction of the people who work for the organisation, by offering opportunities for participation, responsibility, team working etc.

- Support value-adding, customer-focused business processes.

- Support and improve organisational performance through all of the above.

1.1 Aspects of organisation structure

Some of the decisions that will have to be made in designing or evaluating an organisation structure according to Huczyinski & Buchanan (2001) include:

Specialisation

- How should the work of the organisation be divided up?

- Should tasks be grouped to allow units to specialise (allowing efficient focusing of training, equipment and management).

- Should specialisation be minimised (to simplify communication and allow units and individuals to be versatile and flexible)?

Hierarchy

- Should the overall structure be 'tall' (with many levels or tiers of management) or 'flat' (fewer tiers – meaning that each manager has to control more people: a wider 'span of control')?

- Tall organisations allow closer managerial control (due to narrower spans of control), while flat organisations save on managerial costs and empower and satisfy workers (who are given more authority and discretion).

Grouping

- How should jobs and departments be grouped together?

- Options include departmentation by the specialist expertise and resources required (function), or the services/products offered (product/market/brand), or the geographical area being targeted (territory).

Co-ordination

- How can the organisation foster integration between its different units, in order to maximise the 'horizontal' flow of business processes from one to the other?

- This will involve mechanisms such as rules and policies, carefully aligned goals and plans, liaison/co-ordinator roles, cross-functional team-working and so on.

Control

- Should decisions be mainly 'centralised' (taken at the top) or 'decentralised' delegated to lower levels) – or a mixture of both?

- Centralisation allows swift, decisive control and co-ordination, while decentralisation empowers and satisfies workers, and can support organisational responsiveness by having decisions taken closer to the customer or local market.

1.2 Elements of organisation structure

Henry Mintzberg (1984) provided a framework and language for discussing organisation structure, by categorising the building blocks of organisation. He suggested that all organisations can be analysed into five components, according to how they relate to the work of the organisation.

The components can be explained as follows.

Component	Features
Strategic apex	Ensures the organisation follows its mission and services the needs of its key stakeholders.
	Manages the organisation's relationship with the environment (boundary management).
	Acts as a force for direction (shared vision and goals).

Component	Features
Operating core	People directly involved in the process of obtaining inputs, and converting them into outputs (goods and services). Acts as a force for proficiency (competence). Conveys the goals set by the strategic apex and controls the work of the operating core in pursuit of those goals: ie middle management. Acts as a force for concentration (technical or product specialisation and accountability).
Technostructure	Analyses, determines and standardises work processes, techniques, skills and outputs. Examples include strategic planners, quality controllers, human resource management: specialist advice and analysis. Acts as a force for efficiency.
Support staff	Ancillary and administrative services such as PR, legal counsel, building maintenance and security. Support staff do not plan or standardise production, but function independently of the operating core. Acts as a force for learning.

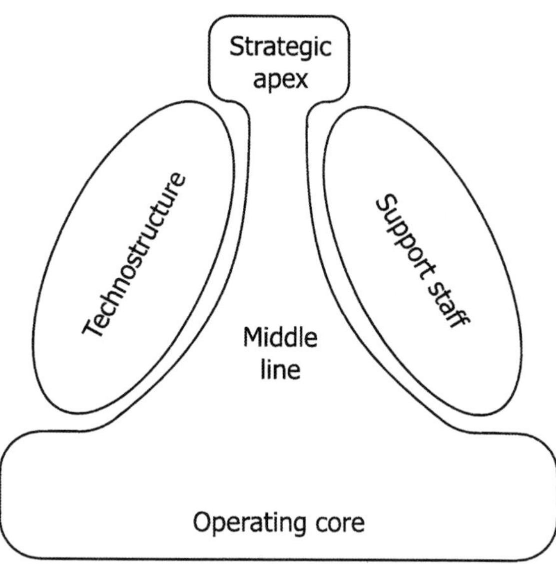

Mintzberg's organisational components

These various components serve to co-ordinate the activities of the organisation in different ways:

- **Direct supervision**: favoured by the strategic apex and middle line. Power is directly applied to control and co-ordinate activity.

- **Mutual adjustment**: the integration of goals by communication and negotiation, favoured by the operating core and support staff.

- **Standardisation** of work processes, outputs and/or skills and knowledge, favoured by the technostructure.

The organisation also has a sixth component, which Mintzberg calls ideology: its paradigm or set of guiding assumptions and beliefs. This is identified with organisation culture, which we discussed in Chapter 3 of this Study Text.

Mintzberg's (1983) components are particularly useful when discussing issues of organisational structure. The vocabulary is precise and well-established.

By configuring these components in various ways, Mintzberg thus divides organisations into five broad types.

Simple (or entrepreneurial) structures

- Small, hierarchical organisations based on centralised control by a single leader Because of their small size and strong hands-on leadership, they are characterised by coherent direction, informal relationships and flexibility.

- This structure is suited to small, entrepreneurial owner-managed firms.

- It consists mainly of the strategic apex and operating core.

Machine structure

- Hierarchical, bureaucratic organisations, suited to stable environments and tasks requiring strict standardisation and compliance.

- This structure tends to have many layers of middle line, and an enlarged technostructure (to standardise work procedures) and staff support.

Professional structure

- Such as an accountancy or legal practice.

- Hierarchical, but recognising the expectations and abilities of professional staff, and therefore flatter and more participative.

- This structure has little middle line or technostructure (since the operating core is comprised of experts), but a substantial staff support element for administrative and clerical support.

Divisional structure

- A number of more or less autonomous divisions, co-ordinated by centralised strategic and support functions.

- This is suitable for devolved structures (such as regionally or internationally dispersed divisions), where central direction is required for brand identity, investment strategy and so on.

- Each division effectively duplicates the whole middle line/operating core/technostructure/support staff structure, while reporting to the centralised strategic apex.

- The centralised technostructure and support structures may appear relatively small.

***Ad hoc* structure**

- An organic, decentralised structure of temporary, flexible project teams and networks.

- This is suitable for new technology (IT, R&D) businesses, consultancies, small media companies and other organisations focused on creativity and innovation.

- The structure tends to mostly technostructure and support staff, flexibly banding together and acting as an operating core where required.

1.3 Influences on organisation structure

Organisational structure is shaped by many factors. There are certain internal principles and dynamics of work organisation such as:

- How far power and authority are held at the top (centralised) or given to lower levels (decentralised).

- The span of control (the number of subordinates that can be supervised by any one superior).

- The division of labour; the grouping of people into working units.

- The need for communication channels, and so on.

These determine some elements of structure, to an extent, according to internal logic.

According to contingency theory, however, there are still managerial choices to be made in order to optimise the structure. A number of contingent variables may influence structural choices and organisational development.

- The strategic objectives or mission of the organisation, and how these are broken down to define and guide the work of sub-units. Diversified organisations, for example, may require more decentralised structures.

- The task or 'business' of the organisation, which will determine which line or task functions are required (development, production, marketing, finance) and which support or staff functions (HR, planning, quality control, maintenance).

- The technology of the task may necessitate certain forms of organisation to maximise its efficiency (eg assembly line organisation of mass production) and the needs of people (eg team working to enhance human involvement in highly automated tasks).

- The size of the organisation. As it gets larger, its structure will get more complex: specialisation, subdivision and formalisation are required in order to control and co-ordinate performance (typically leading to the bureaucratisation of large organisations).

- Geographical dispersion may require federalised structures to take into account relevant factors at local, regional, national, international or global levels of operation.

- The environment of the organisation. Factors (and especially changes) in the political, economic, legal, technological and social environment represent demands and constraints on organisational activity, and opportunities and threats to which organisation structure must adapt.

- The culture and management style of the organisation such as the willingness of management to delegate authority and adopt more fluid facilitate-and-empower roles; organisational values about team working, formality, flexibility and so on.

1.4 Evaluating organisation structures

Drucker (1955) argued that: 'good organisation structure does not by itself produce good performance. But a poor organisation structure makes good performance impossible, no matter how good the individual managers may be. To improve organisation structure....will therefore always improve performance.'

Signs that a structure may be ineffective or dysfunctional include problems such as:

- Slow decision/response times, due to the need to refer decisions via overly lengthy formal communication channels.

- Inter-departmental conflicts, due to ambiguities or overlaps of responsibility.

- Excessive layers of management (often in the middle line), which slows communication, increases overheads, and often requires work creation (or initiative-stifling micro-management) to justify the positions.

- Lack of co-ordination between units, seen in customer complaints, production bottlenecks, inconsistent communications and the proliferation of special co-ordinating mechanisms (liaison officers, committees etc).

- High labour turnover among skilled junior staff, suggesting lack of delegation, challenge and/or development opportunities.

- Lack of identifiable accountabilities for key tasks or results.

1.5 Organisational life phases

In considering organisational structures and designs it is also important to recognise that organisations are required to adapt, reflect and change as they grow and evolve.

Greiner's (1972) organisational growth model provides a framework against which managers can derive a framework within which they can understand their circumstances as a basis for decision-making.

Greiner (1972) model identifies five stages of growth:

Stage	Features
Creativity	An entrepreneurial organisation finds a niche in the market.
	Those involved in the business are product or technology-focused.
	Solution to managerial problems tend to be viewed in terms of better technology and more new products rather than strategic direction and management.
	SMEs can get stuck at this level as the owner/manager fails to put in place the management resource necessary to grow the business.
Direction	Appropriate management structures, financial controls and planning procedures are put in place to enable next phase of growth and development.
	Speed and pace of change means decisions need to be taken closer to where the action occurs.
Delegation	Decision-making becomes decentralised.
	Senior management sets performance criteria and operational management run the business.
	Greater freedom given to managers which can lead to conflicting action impacting on the overall mission of the organisation.
Co-ordination	A better solution than increasing control and regulation is to look for more effective ways of co-ordinating activities.
Collaboration	As the organisation grows co-ordination becomes more difficult to manage.
	Further evolution of the organisation is achieved through such techniques as team building, matrix structures and process orientation.
	While this framework does not provide a description of solutions it helps managers understand the influences upon the organisation and possible routes to a solution.

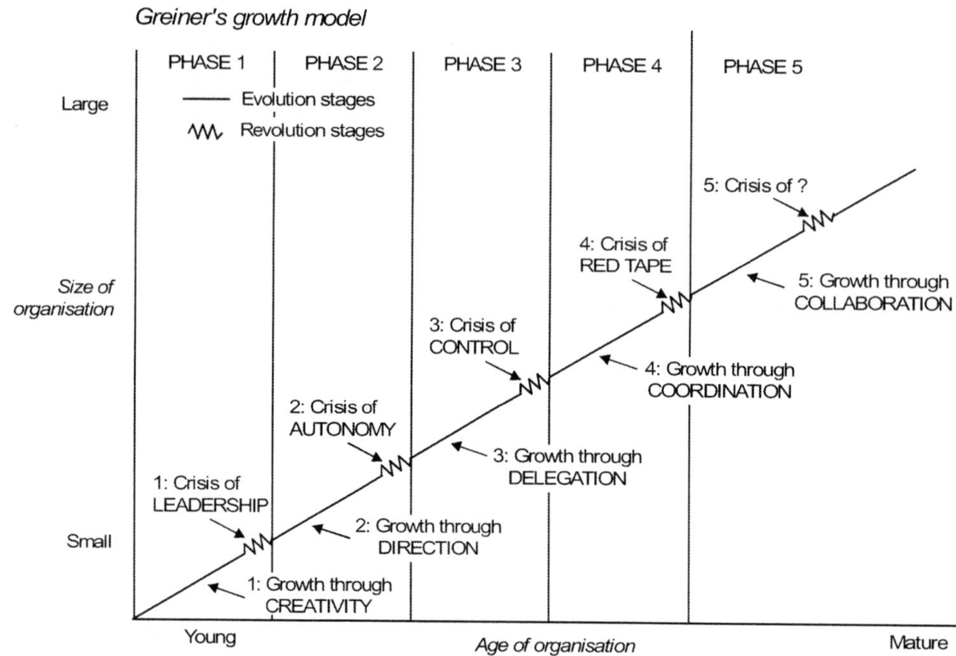

Source: Greiner's growth model (1972)

Activity 1

At what stage in its development is your organisation and what do you recommend needs to be done to move to the next?

1.6 Developing a marketing-oriented organisational structure

Kotler (2002) suggests that as a business grows, it will adopt an increasingly evolved and integrated view of marketing which supports the delivery of marketing value, and the development of marketing focus: that is, an extension of marketing influence throughout the organisation, so that all units cultivate a customer-focused marketing orientation.

Marketing value and focus can be supported in an organisation structure by:

- Positioning the marketing function centrally, at a strategic level, so that strategy flows down from a marketing-oriented strategic apex.

- Business process alignment (or realignment): taking a horizontal process view of the organisation (facilitating the flow of work, information and value towards the customer) and aligning all business processes (research and development; purchasing, supply and distribution; production; sales and marketing) so that they service the same objectives of customer satisfaction.

- Establishing communication mechanisms for internal marketing: supporting the sharing of marketing values and information throughout the organisation.

- Establishing mechanisms for cross-functional collaboration: supporting the involvement and influence of marketing personnel in areas (such as product development and quality management) which potentially impact on customer satisfaction, and therefore benefit from marketing's understanding of customer needs.

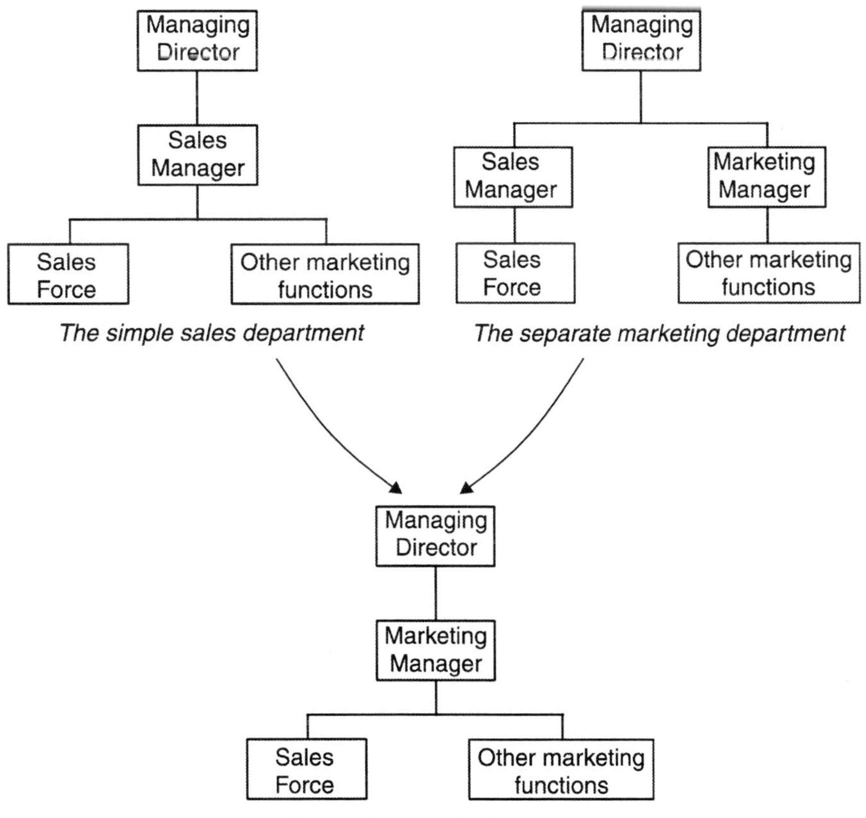

The simple sales department *The separate marketing department*

The modern marketing company

The evolution of the marketing organisation

1.7 Developing an organisational structure that supports creativity

An organisation structure can support creativity and innovation by:

- Encouraging high-frequency multi-directional communication (particularly lateral and upward) for information and ideas-sharing. This is commonly achieved through cross-functional teamworking, meetings (eg for ideas brainstorming) and data-sharing.

- Encouraging flexibility, by removing rigid controls in the form of tight job descriptions, detailed rules and procedures, close supervision, status barriers between managers and staff and so on.

- Structural flexibility supports cultural flexibility, or flexible thinking. Temporary project or task force teams may be formed as and when required for particular tasks, taking advantage of available expertise and ideas.

- All structures and arrangements should ideally be regarded as subject to alteration in response to changing conditions, since their purpose is to facilitate learning, innovation and performance.

- Encouraging initiative, experimentation and risk-taking, by devolving genuine authority to lower levels of staff and removing layers of control, supervision and upward referral. This will have to be supported by actual managerial behaviours as well: for example, not 'punishing' mistakes made while learning or taking initiative, but regarding them as 'learning opportunities'.

- Softening boundaries – not just within the organisation (status boundaries between levels in the hierarchy, and boundaries between functions) but between the organisation and the environment. 'Boundary workers' who have contacts outside the organisation should be encouraged to gather and share information to broaden the knowledge base.

- Opportunities should be sought to learn from other organisations (including customers and competitors) eg by benchmarking and imitation.

These kinds of 'enabling structures' are considered a key characteristic of 'learning organisations' (Pedler, 1991). Organisations which facilitate the acquisition and sharing of knowledge, and the learning of all their members, in order continuously and strategically to transform themselves in response to rapidly changing and uncertain environments.

1.8 Organisation charts

Organisation charts act as a guide to explain how different task roles and positions relate to each other, and how they are co-ordinated and integrated. They typically show:

- **'Shape'**: whether the organisation perceives itself as hierarchical (top-down), or in some other way. A chart may take the form of concentric circles radiating out from a central leader, for example; or a 'web' or network of interrelationships and communication; or an upside-down pyramid with the pointy (senior management) end at the bottom (showing that management sees its role as supporting operational staff in serving the customer, rather than imposing its wishes on them).

- **Hierarchy**: the number of tiers (or circles) of authority in the organisation – and so, depending on the span of control, whether the organisation is tall or flat.

- **Communication channels**: the lines connecting different levels (vertical) and units (horizontal) in the organisation, through which information flows in all directions. Horizontal lines are particularly important, as they reflect cross-functional or inter-group communication: a crucial mechanism of co-ordination.

- **Line, staff and functional authority relationships.** Staff relationships are often shown by a horizontal dotted line between the advisory and client units. Functional authority relationships are shown as a horizontal line joining the vertical chain of command, to show that the unit has authority 'over' aspects of line departments' work.

An organisation chart is only a static 'snapshot' of the planned formal structure of an organisation at a moment in time. It cannot express the quality of management or relationships, or whether information flows effectively via formal channels. It does not attempt to show the true, complex pattern of relationships and communication in the organisation, taking into account informal contacts through 'grapevine'.

An organisation chart may depict authority relationships, but not power or leadership relationships. It may detail formally designated job roles – but not what people actually do, especially if jobs are loosely defined to allow for initiative, growth and flexible contribution.

Nevertheless, drawing an organisation chart may be a helpful way of:

1   Clarifying basic features of the organisational structure.

2   Highlighting potential inefficiencies and dysfunctions, such as: long lines of communication; lack of cross-functional communication; duplications of activity; unclear boundaries of authority or unclear authority relationships.

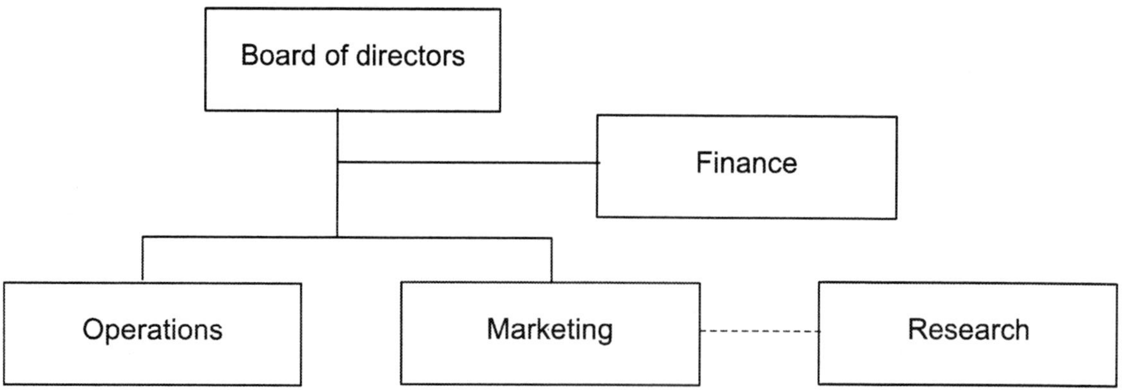

Excerpt from a simple organisation chart

Draw an organisation chart for your own organisation, in any way that expresses, for you, both the structure and the self-perception (shape) of the organisation. Get hold of the formal organisation chart (if any) of your organisation. Compare the two versions and discus with your Study buddy:

- How far the formal organisation chart expresses the self-image and nature of the organisation;
- The effectiveness of the organisation and what potential strengths and weaknesses are highlighted by the chart.

2 Structural frameworks

2.1 Departmental structures

In most organisations, tasks and people are grouped together in some rational way, on the basis of specialisation, say, or shared technology or customer base. This is known as departmentation.

Different patterns of departmentation are possible, and the pattern selected will depend on the goals and dynamics of the organisation.

2.2 Functional (or process-based) structures

Functional organisation involves grouping together people who do similar tasks (and therefore require similar skills, facilities and equipment).

Functions in the organisation as a whole might be production, sales, finance, general administration, marketing.

Subsections of the marketing function might include: communications (or advertising and public relations), market research and sales. Functional specialists such as a market research manager or PR manager will take responsibility for all activities in their specialist discipline across all products and markets.

There are several advantages and disadvantages to such a structure.

Advantages	Disadvantages
Specialist expertise is pooled and focused. Individuals and units can develop their specialisms, while the marketing director co-ordinates their plans and budgets to ensure the development of a coherent and consistent (integrated) marketing mix.	Tendency to focus internally on functional specialisms, processes and inputs, rather than on business processes, the customer and outputs, which are what ultimately drive a business. Inward-looking businesses are less able to adapt to changing demands.
Avoids duplication (eg having a marketing manager for each geographic area or product group).	Lack of 'big picture' awareness of whole-business, value-creating processes.
Enables economies of scale and efficient sharing of equipment and facilities.	Poor co-ordination and communication between functions, due to formal vertical channels of communication and specialist focus (and jargon).

Advantages	Disadvantages
Facilitates the recruitment, management and development of functional specialists.	The creation of vertical barriers to information and work flow.
	Tom Peters (1989) suggests that efficient business processes and effective customer service requires 'horizontal' flow between functions – rather than passing the customer from one functional department to another.
	Loss of control once the organisation's range of products and markets expands: the functional manager may not be able to stay on top of the full range of products, markets or brands, and the burden of co-ordination between functional units may become too great.

The diagram below shows an example of a functional structure.

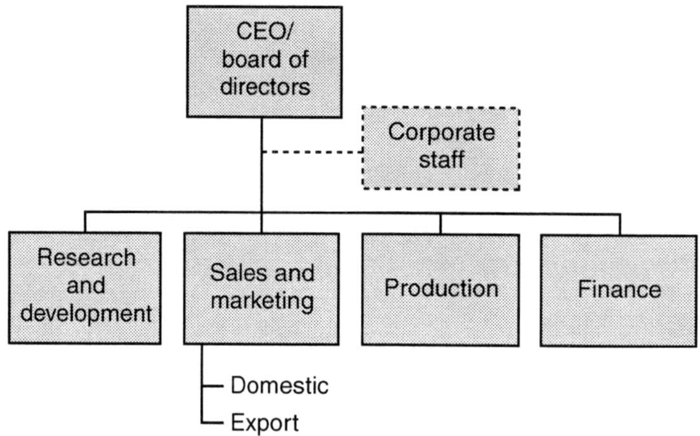

2.3 Product, market and brand structures

2.3.1 Product structure

Some organisations group activities on the basis of products/product groups/'lines'. However, some functional departmentation remains (eg manufacturing, distribution, marketing and sales) but a divisional manager is given responsibility for the product/line, with authority over personnel of different functions.

Product management involves adding an additional tier of management with responsibilities for: developing and implementing marketing plans for particular products or lines; gathering feedback on the products' performance; and initiating product modifications to meet market and competitive pressures.

This type of approach may be particularly appropriate for organisations with a very diverse or very large range of products – especially if some of them are direct or indirect competitors with each other.

Advantages	Disadvantages
Accountability	
Individual managers can be held accountable for the profitability of individual products or lines.	Increased overhead costs and managerial complexity.
Specialisation	
Product managers can build up considerable experience and understanding of their product groups and markets, which is valuable in a rapidly changing competitive environment.	The risk that different product divisions may become competitive, fragmenting objectives, markets and resources in a way that is sub-optimal for the organisation as a whole.

Advantages	Disadvantages
For example, salespeople may be trained to sell a specific product in which they may develop technical expertise and thereby offer a better sales service to customers.	
Co-ordination The different functional activities and efforts required to make and sell each product can be co-ordinated and integrated by the divisional/product manager.	

The product-based approach is becoming increasingly popular, because the benefits of managers having accountability and expertise in specific product areas outweigh the costs associated with a loss of functional specialisation.

The diagram below shows a product-based structure:

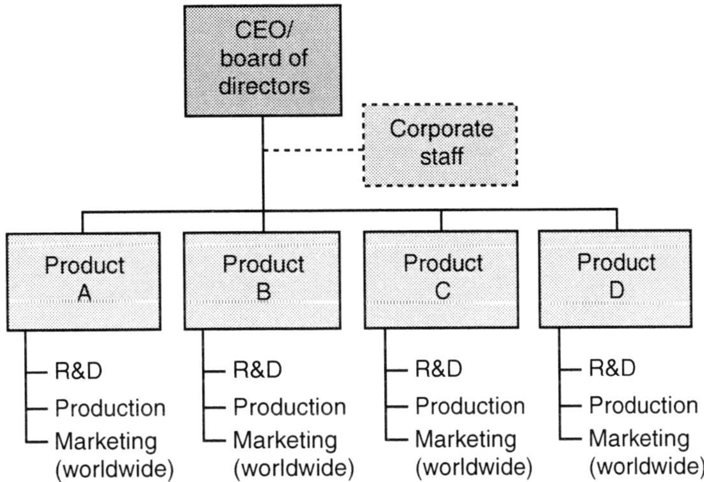

2.3.2 Brand structure

Brand structure is a variant of product structure (common in B2B marketing), focusing managerial authority and accountability on brands rather than product groups: this is common for FMCG producers with a range of potentially competing brands. As with product and market management, some functional organisation remains but brand managers have responsibility for the brand's marketing across all functional activities.

2.3.3 Market or customer type (market management)

In a variant on the product-based approach, managers may take responsibility for particular markets or customer segments.

2.4 Geographical structure

This is an ideal structure if markets vary considerably and beneficial where products are largely homogenous and there is a requirement for fast and efficient distribution. As global sales grow a structure emerges that becomes directly responsible for the development and implementation of the global marketing strategy.

Advantages	Disadvantages
Ability to respond quickly and easily to the environmental and market demands of a region or national area with minor adaptations.	Duplication may result and the challenge is for the organisation to ensure knowledge transfer and product introductions across regions.
Encourages adaptive global marketing programmes.	Inconsistency in methods or standards may develop across different areas.
Economies of scale and cost efficiencies can be achieved.	
Ensures best use of regional expertise and a degree of area autonomy and ownership.	

Within the geographical structure there may be two further structures:

Regional management centres	Where sales volume become high there will be a need of specialist staff to focus on that particular region.
	Appropriate where there is a need to create adaptive programmes across the region.
Country-based subsidiaries	Characterised by a high degree of adaptation to local conditions.
	Each subsidiary develops its own activities and is largely autonomous.

The diagram below shows an example of a geographical structure:

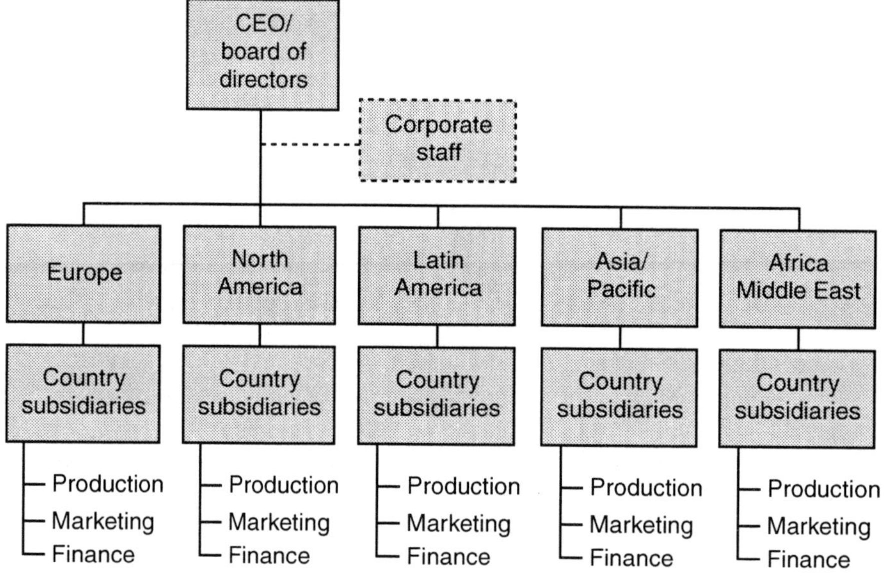

Global case study

EMI Music streamlines global and international marketing structure

EMI Music has reorganised its global and international marketing structure to maximise sales of its artists' repertoire worldwide through all available channels. New global and international marketing teams will manage major frontline releases from priority artists, while a global catalogue marketing function will handle catalogue campaigns across the world.

The EMI Music International division is to form a new international marketing team to focus on the promotion of worldwide priorities within all markets under the division's remit, as well as to identify repertoire and projects from the region with global and multi-region potential. The EMI Music International region comprises Continental Europe, Middle East & Africa, Australasia, Japan, Asia and Latin America.

Jean-Francois Cecillon said: "In this increasingly diversified market, we must ensure that we deliver effective marketing campaigns and products that connect our artists' music with consumers through all available channels. The new teams will be empowered to make effective decisions that will enable us to deliver success for our talented artists on a global basis."

http://www.ifpi.org/content/section_news/20070316b.html

Discussion

Study buddy

A global bank has branches in India, France, Germany and Malaysia. It grew from the merger of a number of small local banks in these countries. These local banks were not large enough to compete single-handedly in their home markets. It hopes to attract both retail and corporate customers, through its use of home banking services and heavily advertised a direct banking service, which is a branchless bank to which customers telephone, e-mail or fax their instructions.

Discuss with your study buddy what sort of organisation structure you think would be appropriate and share your thoughts in the discussion forum.

2.5 Matrix and project-based structures

The idea of the matrix emerged at US aerospace company Lockheed in the 1950s, when its customer (the US government) became frustrated at dealing separately with a number of functional specialists when negotiating defence contracts.

A matrix structure can consist of two organisational structures (eg geographic and product structures) intersecting each other with dual reporting relationships.

Members retain a working relationship within their individual disciplines at the same time as reporting to a project leader for the specific project in hand.

Matrix structures have been implemented in manufacturing and service organisations in an attempt to enhance the integration of functional specialisation with more effective design, manufacturing and marketing of products and services.

Advantages	Disadvantages
A dual focus ensures that concerns can be identified and analysed objectively.	It may lead to tensions between divisions, between global and local demands and issues.
It is most useful for organisations that are both product diversified and geographically spread.	An individual with two or more bosses may suffer stress from conflicting demands or ambiguous roles.
Combining product management and market-oriented approach ensures the organisation can best meet needs of both markets and products.	Slower decision-making due to the added complexity.
Employees develop an attitude geared to accepting change, and departmental monopolies are broken down.	
Inter-disciplinary co-operation and a mixing of skills and expertise, along with improved communication and co-ordination.	
Provides employees with greater participation in planning and control decisions.	

Advantages	Disadvantages
The organisation tends to become more customer/quality focused.	
Bureaucratic obstacles are removed, and departmental specialisms become less powerful.	

The diagram below illustrates a matrix structure.

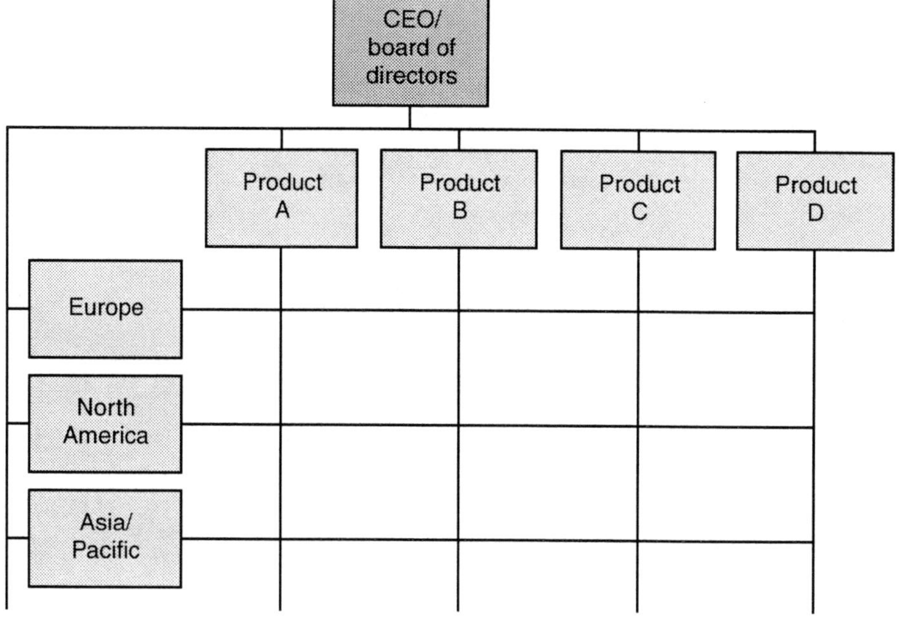

How might a matrix structure be appropriate in your organisation? What will be the advantages and what will be the issues that need to be considered?

2.6 The mechanistic and organic organisation

The terms 'mechanistic and 'organic' were coined by Burns and Stalker (1961) to describe forms of organisation which are:

- Stable, efficient and suitable for slow-changing operating environments (mechanistic or 'machine-like' organisations also called 'bureaucracies').

- Flexible, adaptive and suitable for fast-changing or dynamic operating environments (organic or 'organism-like' organisations).

The characteristics of these organisations can be summarised as follows.

Factor	Mechanistic	Organic
The job	Tasks are **specialised** and broken down into sub-tasks.	Specialist knowledge and expertise is understood to contribute to the **common task** of the organisation.
How the job fits in	People are concerned with completing the task **efficiently**, rather than how the task can be made to improve organisational **effectiveness**.	Each task is seen and understood to be set by the **total situation** of the firm: focus is on the task's contribution to **organisational effectiveness**.

Aligning the organisation to the vision

Factor	Mechanistic	Organic
Co-ordination	**Managers** are responsible for co-ordinating tasks.	People adjust and redefine their tasks through interaction and **mutual adjustment** with others.
Job description	There are **precise** job descriptions and delineations of responsibility.	Job descriptions are **less precise**: people do what is necessary to complete the task.
Commitment	**Doing the job** takes priority over serving the interests of the organisation.	**Commitment to the organisation** spreads beyond any technical definition of competence.
Legal contract v common interest	**Hierarchical** structure of control. An individual's performance and conduct derive from a **contractual relationship** with an impersonal organisation.	**Network structure** of control. An individual's performance and conduct derive from a supposed **community of interest** between the individual and the organisation, and the individual's colleagues.
Decisions	Decisions are taken by **senior managers** who are assumed to know everything.	Relevant technical and commercial knowledge can be located **anywhere**.
Communication patterns	Communication is mainly **vertical** (up and down the scalar chain), and takes the form of **commands** and obedience.	Communication is **lateral** or networked. Communication between people of different rank represents **consultation**, rather than command.
Content of communications	Operations and working behaviour are governed by **instructions** issued by superiors.	Communication consists of **information and advice** rather than instructions and decisions.
Mission	Insistence on **loyalty** to the concern and **obedience** to superiors.	Commitment to the organisation's **mission** is more highly valued than loyalty, as such.
Internal v external expertise	**Internal knowledge** (eg of the organisation's specific activities) is more highly valued than general knowledge.	Importance and prestige attach to affiliations and expertise valid in the industrial, technical and commercial milieux **external** to the firm.

Burns and Stalker (1961) argue that while bureaucracy can be highly efficient in stable, slow-change environments, which do not require sensitivity to customer or external environmental demands, organic structures are better suited to conditions of change – particularly where there is a need to respond to change by continuous innovation and creativity, as is arguably the case in most marketing environments.

Organic organisations are typified by:

- A 'contributive' culture of information and skill sharing, encouraging versatility (rather than specialisation) and team working (rather than functional departmentation).

- A 'network' structure of authority and communication, allowing decentralisation and a range of lateral relationships (crossing functional boundaries) for co-ordination and self-control.

- Focus on goals and outputs rather than processes.

- Job design that allows flexible definition of tasks according to the needs of the team and changing demands.

Note that the two approaches represent two ends of a spectrum: there are intermediate stages between bureaucratic and organic organisations.

Different departments of a business may be run on different lines. For example, the finance function has a well-defined task with little variation: controls are needed to ensure processing accuracy and to avoid fraud. A mechanistic system might be applied here.

On the other hand, the marketing function, employing professional experts and creatives, requiring sensitivity to environmental changes and customer trends, and needing to network widely within and outside the organisation, may be run on an organic basis.

2.7 Centralisation and decentralisation

A major concern of management when considering organisational structure is division of labour or specialisation (ie who does what, where does responsibility lie, what authority/control over resources do individuals have).

2.7.1 Factors affecting the extent of decentralisation

Organisation type chosen	The organisation structure types, previously discussed, vary in the extent to which they lend themselves to a decentralised structure.
The nature of the management function	Marketing tends to be a decentralised function, compared with finance for example, because of the importance of understanding local customer needs, responding to environmental differences and changes, developing relationships with customers and so on.
The size of subsidiaries	The larger a subsidiary is the more autonomy it will tend to have in its decision-making, (ie the more decentralised it will be).

2.7.2 Advantages and disadvantages of centralisation

Advantages of centralisation	Better planning (providing that 'top-down' planning is indeed the most effective). Better co-ordination in achieving overall organisational objectives. Optimum use of resources (operational, financial and human).
Disadvantages of centralisation	Motivation of senior staff at subsidiary level may be adversely affected if they do not have control over decisions or resources which affect their targets. Centralisation may lead to misunderstandings and delays in management communications.

Activity 3

For your organisation consider the advantages and disadvantages of a centralised v decentralised organisational structure and which you feel is most appropriate.

3 Alternative organisations

3.1 Recent trends

Some recent trends have emerged from the focus on responsiveness and flexibility as key organisational values.

- **Flat structures**
  The flattening of hierarchies does away with levels of organisation which lengthened lines of communication and decision-making and encouraged ever-increasing specialisation. Flat structures are more responsive, because there is a more direct relationship between the organisation's strategic centre and the operational units serving the customer.

- **Horizontal structures**

  What Peters (1989) calls 'going horizontal' is a recognition that functional versatility (through multi-functional project teams and multi-skilling, for example) is the key to flexibility. In the words (quoted by Peters) of a Motorola executive: 'The traditional job descriptions were barriers. We needed an organisation soft enough between the organisational disciplines so that ... people would run freely across functional barriers or organisational barriers with the common goal of getting the job done, rather than just making certain that their specific part of the job was completed.'

- **'Chunked' and 'unglued' structures**

  This typically means team working and decentralisation, or empowerment, creating smaller and more flexible units within the overall structure. Charles Handy's 'shamrock organisation' (with a three-leafed structure of professional core, contractual fringe and flexible labour force) is gaining ground as a workable model for a leaner and more flexible workforce, within a controlled framework (Handy, 1989).

Each category is managed and rewarded differently:

- **Professional core**

  Professional workers who direct the organisation in its technical and professional endeavours.

- **Contractual fringe**

  Subcontractors of various descriptions who provide necessary, but not essential, services.

- **Flexible labour force**

  Part time, casual and freelance workers that come and go as needed.

The purpose of the model is to encourage managers to consider organisational purpose together with the nature of the staff relationship to the organisation. There are organisational structure implications in that the three categories of worker engage with the organisation differently and therefore need to be integrated and managed differently.

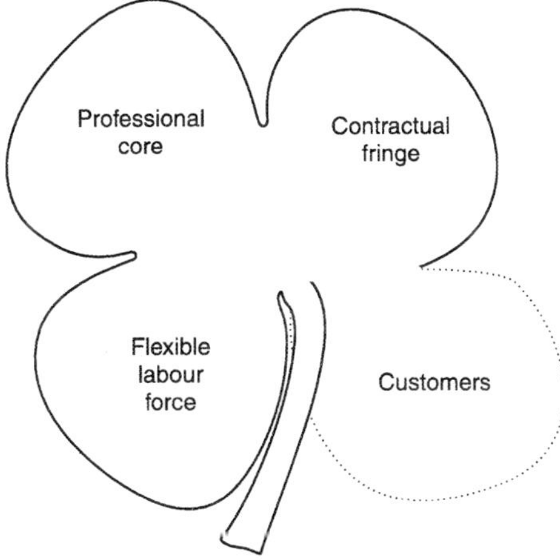

The shamrock organisation

- **Output-focused structures**

  The key to all the above trends is the focus on results, and on the customer, instead of internal processes and functions for their own sake. A project management orientation and structure, for example, is being applied to the supply of services within the organisation (to internal customers) as well as to the external market, in order to facilitate listening and responding to customer demands.

- **'Jobless' structures**

  The employee becomes not a job-holder but the vendor of a portfolio of demonstrated outputs and competencies (Bridges, 1994). However daunting, this is a concrete expression of the concept of employability, which says that a person needs to have a portfolio of skills which are valuable on the open labour market: employees need to be mobile, moving between organisations rather than settling in to a particular job.

- **Boundaryless structures**

  Milkovich & Brodie (1997) state that horizontal (status) barriers are removed or softened by delayering and participative decision-making, and vertical (functional) barriers by empowered cross-functional team working – in order to align business processes. The boundaries of the organisation are also softened, by co-opting suppliers, distributors, business allies and customers as collaborators in the value-adding process, creating an 'extended' or 'networked' enterprise.

- **Adhocracy**

  Huzcynski & Buchanan (2001) state that this is '*a type of organisation design which is temporary, adaptive, creative – in contrast to bureaucracy, which tends to be permanent, rule-driven and inflexible*'. It is associated with creative thinking, innovation and organisational learning. It would typically involve the use of a loose network of flexible, temporary, cross-functional project teams or networks of specialists, banding together and disbanding as required to seize business opportunities. While ideally suited to turbulent, innovative markets, it presents a challenge for managers and employees, since it means living with disorder, ambiguity and uncertainty on an on-going basis.

Activity 4

Consider how elements of such structures could be effectively built into your organisation and the advantages of doing so.

---

4 Factors influencing organisational design

The choice of organisational design and structure will depend partly on the preferences of top management, and the style they like to adopt, but there are other influences as follows.

- In terms of the marketing mix, the element most likely to be centrally determined is product. Distribution channel decisions may be shared between head office and local management. Advertising and promotion decisions may need to be made at a local level for linguistic and cultural reasons.

- When big capital investments are planned, head office should be involved in the decision.

- When cash flow is tight, other strategies must be sacrificed to the paramount concern for short-term survival and attention to cash flow.

- Widely diversified organisations, and organisations in industries which are facing rapid change, would be better managed by a greater decentralisation of authority.

- Organisations in a single industry which is fairly stable would perhaps be more efficiently managed by a hierarchical, centralised management system.

There are a number of departments, functions, activities and components that need to be fitted together within a cohesive framework. It is necessary to integrate all these components in order that the vision can be realised effectively.

Key processes must also be integrated together with the organisational structure such as financial reporting systems, communication, consultative and reporting mechanisms, organisational policies and so on.

An organisational structure can be weakened through poor support infrastructure but any weaknesses can be offset by the support infrastructure in place.

Child (1972) suggested six components necessary for the creation of an organisational structure:

Component		Features
1	Task	Reflects the way actual tasks need to be performed.
2	Reporting relationships	Reflects the way managers will be imposed on the organisation in a girded hierarchy of responsibility.
3	Clustering	Reflects how groups of activity are clustered together in a recognisable pattern.
4	Procedures	Relicts systems, policies, procedures, communication networks and information flow that impact on integration and co-ordination with stakeholders.
5	Delegation	Reflects decision-making discretion at the different levels in the graded hierarchy.
6	Motivation	Impacts on the number of people needed to perform the tasks, style of management and relationships between employees and management.

GMN viewpoint

In his book, 'The Self Destructive Habits of Good Companies', GMN Global Honorary President Dr Jagdish Sheth warns against the following complacencies in good companies relating to organisational structure and design:

Processes are overly bureaucratic

Decision-making is thwarted by committees with cultural differences between the members that are irreconcilable. It is the leader's responsibility to break down the bureaucracy and get rid of other obstacles that are hallmarks of no-hurry complacent organisations.

Organisation structure is complete vertical integration

Everything is done internally even though it will be cheaper and more cost effective to outsource.

Sheth makes a number of suggestions:

1. Re-engineer the organisation and look for where processes can be improved, efficiencies can be gained and so on.

2. Reorganise around delivering most value at lowest total cost to the market. For example, Hewlett Packard recognised it had one worldwide market and reorganised around global product management responsibilities.

Reorganising can facilitate and improve quality and service levels when it is embedded in a process of change deliberately designed to raise quality standards and stakeholder satisfaction.

5 Quality concepts

Product and service quality are critical success factors for marketing-oriented businesses, because they determine customer satisfaction – the ultimate source of competitive survival and advantage. The ability of a company to differentiate its products and services in the marketplace substantially rests on the quality of its total offering: its ability consistently not just to satisfy, but to delight – and hence retain, and potentially gain the loyalty of – its customers. So what is 'quality'? What makes a 'quality product' or offering?

5.1 Quality defined

> **Definition**
>
> **Quality** is a measure of the features and characteristics of a product or service which affects the ability to meet stated or implied needs.

Definitions of quality have focused on dimensions such as:

1   **Quality of design**: the potential customer satisfactions built into a product.

2   **Fitness for use or fitness for purpose**: that is, the extent to which a product does what it is intended and expected to do, and meets the customer's needs. (This is sometimes called the 'user-based' approach to quality.)

3   **Conformance to requirement** (or **meeting specification**): that is, the product complies with the features, attributes, performance and other aspects set out as a requirement in the product specification. (This is the 'product-based' approach.)

4   **Acceptable quality and value for money**: customers may be willing to sacrifice some performance and features for a lower price, as long as fitness for use is still acceptable. (This is the 'value-based' approach.)

5.2 Dimensions of quality

Garvin (1987) identifies eight generic dimensions of product quality

1   **Performance**: the operating characteristics of the product

2   **Features**: value-adding characteristics and service elements (eg warranties, after-sales service)

3   **Reliability**: the ability of the product to perform consistently over time

4   **Durability**: the length of time a product will last without deterioration

5   **Conformance**: whether agreed specifications and standards are met

6   **Serviceability**: the ease and availability of service support

7   **Aesthetics**: customer perceptions of how pleasing the product looks, sounds, tastes and so on

8   **Perceived quality**: the subjective expectations and perceptions developed by customers, as a result of marketing, brand identity, price and so on

Essentially and inevitably, however, quality means different things to different operations. It is also important to note that customers may perceive quality in different ways – and that quality must always be understood from the customer's perspective! Marketers need to identify 'quality gaps': shortfalls between what customers expect or perceive, and what the organisation provides. Then and only then will the organisation be aligned to the vision.

Slack *et al* (2004) suggest that there may be a gap between the customer's specification and the organisation's specification; between the concept and the specification; between quality specification and actual quality; or between actual quality and the quality promised to the customer. Different strategies may be necessary to reduce or eliminate different gaps, embracing product specification and development, materials and production quality – and marketing.

5.3 The views of some quality 'gurus'

A number of writers have been influential on quality thinking in recent decades.

Juran (1998) argues that quality should not be seen merely in terms of meeting specification (a measure of conformance from the producer's point of view), but should focus on the needs of internal and external customers.

He advocates a user-based approach, focused on fitness for use or fitness for purpose. He emphasises the importance of management's role in quality improvement and the need to motivate the work force and involve them in quality improvement initiatives. He also emphasises that quality management should aim to ensure that the way in which work is performed (ie systems and processes) facilitate high quality output. Juran believes that 85% of quality problems were the result of ineffective systems.

Deming (1986) states that the only judge of quality is, ultimately, the customer. Quality means finding out what the customer needs, wants and values – and providing it, with the aim of meeting customer expectations (customer satisfaction) or going further and exceeding customer expectations (customer delight). Deming is credited with developing the Total Quality Management approach where emphasis shifted to quality as a **managerial philosophy** and underlying organisation-wide systems. Deming also believes that as process variability (the amount of unpredictability in a process) decreases, quality and productivity increase: quality can therefore be improved by reducing process variability. He stresses the need for statistical control methods, participation, education, openness and improvement.

Ishikawa (1991) stresses the importance of people and participation in the process of solving quality problems. He devised the idea of **quality circles** to achieve participation and overcome resistance to quality control, which workers tend to dislike because of its emphasis on statistics and rigid standards. He also introduced the use of 'fishbone' diagrams to diagnose quality problems.

Taguchi (1986) argues that quality issues begin at the product design stage, and promotes the idea of team meetings between managers and workers to discuss product design. He also suggests that it is important for managers to understand the true costs of poor quality, to the organisation and to society, including factors such as warranty costs, customer complaints and the loss of customer goodwill.

Crosby (1979) likewise focuses on the costs of quality, the need for worker participation and the need to motivate individuals to improve quality. His 'absolutes of quality management' are:

- Quality as conformance to requirements
- Effort directed at prevention rather than detection of defects
- Zero defects in production
- Systematic measurement of the cost or price of non-conformance
- No such thing as a 'quality problem'.

6 Quality systems and processes

6.1 Why is it important to maintain quality?

The business case for quality management can be summarised as follows.

Cost: as we mentioned above, the potential costs and risks of quality failure are more extensive, and higher, than the controllable costs of managing quality.

Competitive advantage: quality is a key dimension on which products can be meaningfully differentiated from their competitors and the ability to offer consistently high quality may be a core distinguishing competence for an organisation. At the micro-level, quality comparison is likely to be a key decision factor for a customer choosing between competing products, and quality failures may be sufficient to cause brand switching.

Brand positioning: quality is one of the main dimensions on which a marketing organisation may seek to position a brand in the perception of the market. High quality, premium or luxury brands deliberate engage customers on a brand platform focused on quality attributes. Quality failures may significantly damage the identity and positioning of such brands.

Relationship marketing: quality is a tool of customer acquisition, but the consistent quality of goods and service encounters is also extremely important in customer retention, the building and maintenance of trust, and the engagement of customer loyalty – the main focus of relationship marketing. Repeated disappointments (however well managed) will ultimately be fatal to the maintenance of on-going customer relationships – and reduce customer lifetime profitability.

Price: quality is related to customer perceptions of value for money. The market will often bear higher or premium pricing for products and services perceived to be (or marketed as) of high quality.

Compliance: in some areas, quality is an issue of legal compliance. If goods are not of satisfactory quality or are not fit for their purpose, or are actually or potentially unsafe, the supplier may be liable in law.

Reputation management: quality – and more particularly, safety – failures can be highly damaging to the organisation's corporate reputation and standing, if goods are exposed as 'shoddy' or are forced to be recalled for safety reasons.

Culture and morale: a culture built on values such as customer satisfaction, excellence and quality (whether 'premium/prestige' or 'value for money') is more compelling and positive than a culture of indifference, customer exploitation and 'corner cutting'. Pride in the organisation and brand can be a strong source of employee morale and team spirit.

Employee satisfaction and loyalty: employee satisfaction and loyalty are linked to product and service quality in complex ways. The loyalty-based cycle of growth (Reichheld, 1996) suggests that superior quality and service creates satisfied and loyal customers – but, at the same time satisfied and loyal customers generate revenue which can be invested in employees (eg through superior rewards). Consistent delivery of superior value to customers gives employees pride and satisfaction in their work (and avoids the stress and frustration of complaint handling etc).

6.2 Quality management strategies

Definition

Quality management is concerned with controlling activities with the aim of ensuring that products or services are fit for their purpose, and meet pre-determined specifications. Quality management encompasses a variety of approaches, including quality assurance and quality control.

6.3 Quality management as a business philosophy

In recent decades, and following the one-time competitive success of Japanese management techniques, quality management has come to represent not just a managerial function, but a managerial orientation – affecting all areas of the organisation's structures and processes.

Tom Peters (1989) describes this as a 'quality revolution', characterised by the adoption of quality as a guiding system or ideology for an organisation's culture and systems – and the integration and involvement of the whole (internal and external) value chain on on-going continuous improvement. Perhaps the key embodiment of this orientation is Total Quality Management (TQM).

Definition

A **quality management system** (QMS) can be defined as a set of co-ordinated activities which will direct and control an organisation in order to continually improve the effectiveness and efficiency of its processes and performance. The main thrust of a QMS is defining and managing the processes which will result in the production of quality products and services (a quality assurance strategy) – rather than in detecting defective products or services (a quality control strategy).

A fully documented, systematic QMS is designed to ensure that:

- Customers' requirements are met, and they have confidence in the ability of the organisation to consistently deliver products and services which meet their needs and expectations.

- Organisational performance objectives are consistently achieved, through improved process control and reduced wastage, with the efficient use of available resources (materials, human, technology and information).

- Staff competence, training and morale are enhanced, through clear expectations and process requirements.

- Quality gains are maintained once achieved: learning and good practices do not 'slip' for lack of documentation, adoption and consistency.

7 Total Quality Management (TQM)

Definition

Total Quality Management (TQM) is an orientation to quality in which quality values and aspirations are applied to the management of all resources and relationships within the firm and throughout the value chain, in order to seek continuous improvement and excellence in all aspects of performance.

7.1 Philosophy and values of TQM

Mullins (1999) synthesises various writers' definitions of TQM as expressing: *'a way of life for an organisation as a whole, committed to total customer satisfaction through a continuous process of improvement, and the contribution and involvement of people'.*

The seven key principles and values of TQM includes:

1 **'Getting it right first time'**, namely that quality should be designed into products, services and processes, with the aim of achieving zero defects.

   The traditional approach to quality management argues that there is an optimal level of quality effort which minimises total quality costs: there is a point beyond which spending more on quality yields a benefit that is less than the additional cost incurred. TQM, however, argues that:

   - Failure and poor quality are simply unacceptable: the inevitability of error is not something that an organisation should accept, as a cultural value. The target should be zero defects.

   - Taking into account the 'wider' costs of poor quality – the potential to lose customers, the taking up of managerial time on non-value-adding problem-solving, the loss of morale and trust among staff – no amount of defects can be considered 'optimal'.

- Quality can and should be continually improved, and this need not incur high costs or diminishing returns.

2 Quality chains

- The quality chain extends from suppliers through to consumers, via the 'internal supply chain' (supplier and customer units representing the flow of work within the organisation).

- The work of each link in this chain impacts on the next one, and will eventually affect the quality provided to the consumer.

- The chain may be broken at any point where people or processes fail to fulfil the quality requirements of their immediate (internal or external) customer.

- Clear customer relationships and accountabilities (including service level agreements, where required) need to be established throughout the chain.

3 Quality culture and total involvement

- Quality is a 'way of life': a key cultural value in the organisation.

- Every person within an organisation potentially has an impact on quality, and it is the responsibility of everyone to get quality right.

- This needs to be consistently and compellingly expressed and modelled by senior management, and supported and reinforced by recruitment, training, team building, appraisal and reward systems.

4 Quality through people

- Commitment, communication, awareness and problem-solving are more important in securing quality than mere systems.

- Team-based management and empowerment are key elements of TQM.

- Teams must be empowered and equipped to take action necessary to correct problems, propose and implement improvements, and respond flexibly and fast to customer needs (particularly in service contexts).

- There must be high-quality, multi-directional communication. Strict control systems (and associated 'blaming') should be replaced by a culture of learning, in which people are not afraid to try new things or take the initiative to make small improvements.

5 Process alignment

- Business processes should be deliberately designed and modified so that every activity is geared to the same end: meeting the customer's wants and needs.

- Where this is not the case, there may be the need for radical change programmes such as Business Process Re-engineering (BPR).

6 Quality management

- All processes and operations are monitored and controlled according to clear, specific, measurable performance criteria.

- Attention is focused on getting the process right – and only secondarily on outputs.

- Refusal to focus on short-term effects ensures a more proactive commitment to long-term improvement.

- Quality systems should be thoroughly documented in:

    – A company quality manual, summarising the quality management policy and system;

    – A procedures manual setting out the functions, structures and responsibilities for quality in each department;

–   Detailed work instructions and specifications for how work should be carried out in order to achieve the desired quality standards. These will have to be continually updated as improvements and adjustments are made.

7   **Continuous improvement**

- Quality is not a 'one-off' exercise.

- By seeking to continually improve, organisations remain sensitive to new opportunities and approaches, and encourage learning and flexibility at all levels.

- In contrast to radical, 'discontinuous' or 'blank slate' change approaches such as BPR, continuous improvement may operate by small-step or incremental changes. Such evolutionary change is less traumatic, less risky, less costly, and can be driven from the bottom-up by empowered employees.

7.2 Continuous improvement

Quality management involves the on-going and continual examination and improvement of existing processes: 'getting it more right, next time'.

This process is sometimes referred by its Japanese name of 'kaizen': 'a Japanese concept of a total quality approach based on continual evolutionary change with considerable responsibility given to employees within certain fixed boundaries' (Mullins, 1999).

Kaizen looks for uninterrupted, ongoing incremental change: there is always room for improvement and continuously trying to become better, for example through elimination of wastes (non-value-adding activities), immediately accessible improvements to equipment, materials or team behaviour.

It is essentially a cyclical approach to change, because it incorporates reflection or evaluation for future learning and improvement. It is also essentially a bottom-up or empowerment-based approach to quality, because it utilises the feedback, ideas and initiatives of those closet to quality issues: operational staff and customers.

A basic cyclical approach to kaizen involves:

- Identifying potential areas for improvement, by monitoring performance and gathering feedback from employees and other stakeholders.

- Analysing the data to identify possible causes of common factors prevalent in performance shortfalls or 'hidden messages' in the results.

- Planning and implementing action to improve performance in the identified area.

- Assessing the effects of the action and identifying any further problems (in a continuous cycle).

7.3 Quality circles

Japanese quality guru Ishikawa (1991) stresses the importance of people and participation in the process of solving quality problems. He devised the idea of quality circles to achieve participation, harness the expertise and commitment of staff, share best practice, and overcome resistance to quality management.

Definition

A **quality circle** is a team of workers from different levels and functions of an organisation which meets regularly to discuss issues relating to quality, and to recommend improvements.

A quality circle typically consists of a voluntary-participative group of about eight employees, from different functions, which meets regularly (during working hours) to discuss problems of quality and quality control in their areas of work, and to suggest quality improvements.

The circle is facilitated by a leader who directs the discussion and helps to orient and develop members of the circle (if required) in quality management and problem-solving techniques.

Management should not impose agendas on the meetings, in order to permit openness about problems and conflicts, and to encourage thinking 'outside the box'.

Feedback on suggestions offered and problems raised should, however, be swift and constructive, in order to demonstrate that the quality circle is taken seriously as a participative and quality tool.

7.3.1 Benefits of quality circles

In practice, quality circles may or may not have significant responsibility for making, implementing or monitoring the progress of their recommendations. Even as discussion groups, however, they may have significant benefits, especially if members return to their departments as 'ambassadors' for quality, having been involved in the circle: this may be seen as a tool of internal marketing.

Benefits claimed for quality circles include:

- Greater motivation, satisfaction and commitment of employees, through involvement in key organisational values and processes.

- Improved productivity and quality of output, through involving operational staff, generating new ideas and sharing best practice.

- More multi-directional communication and the establishment of informal (but work-focused) information-sharing networks.

- Development of employees in team working, project management and problem-solving skills.

- Greater awareness of quality and service issues, and customer focus, supporting a 'culture of quality'.

Drawbacks may include:

- Wasted time and cost (if the meetings do not result in applied improvements).

- Perceived powerlessness causing resentment and resistance to other attempts at participation.

- Possible empire building by the quality circle.

- The risk that business requirements (eg cost control) may not be fully understood by the group in making recommendations.

However, quality circles are generally considered constructive, and have expanded to include quality groups drawn from a wider range of stakeholders in the value network (eg including suppliers, distributors and customers) – and best practice sharing groups or networks, which may include a wide range of organisations with a common interest in quality issues.

7.4 Limitations and criticisms of TQM

TQM is susceptible to various adverse perceptions:

- TQM initiatives may not be introduced or implemented effectively, and the job is 'botched' by management.

- After obtaining short-term benefits from introducing TQM, the benefits wear off over time, due to 'quality disillusionment'.

- The introductory phase may be resisted as a temporary 'management fad'.

- The implementation phase may give rise to the perception that TQM is driven and owned by a particular group, rather than a total involvement concept (unless staff can be involved in implementation from the vision stage).

TQM programmes can also suffer from:

- A 'blitz' approach to implementation: introducing TQM rapidly, with mass education and communication programmes. This can lead to problems of not knowing what to do or what to do next, and can disrupt and damage business continuity. (A more gradual approach, gaining stakeholder buy-in and collaboration at each stage, and integrating progressive measures into strategic and functional plans may be easier to implement.)

- Time, cost and difficulty of introducing, implementing and embedding the approach (particularly in large, bureaucratic organisations). The impact of TQM cannot be underestimated, as it represents a major strategic and cultural change.

- A lack of long-term, cohesive and visionary top-management commitment.

- A failure to understand the full range of quality issues and quality costs that should be embraced by quality management systems.

- Vested interests and organisation politics, resisting change attempts. Individuals need to understand the TQM process, its purpose and why it is important for the organisation. They also need to be empowered – and this may cause resistance from managers who fear a loss of control.

- Embedded cultural cynicism about quality and fulfilling customer needs: TQM is difficult to undertake if the organisation merely wants to pay 'lip service' to the concept.

Activity 5

To compete successfully, companies must eliminate weaknesses which may include:

- Doing what has always been done
- Not understanding competitive positioning
- Compartmentalising of operational functions
- Trying to control people through systems
- Confusing quality with grade, or grade with quality
- Having an acceptable quality level
- Fire fighting is regarded as 'macho'
- The 'not my problem' syndrome

Explain each of these weaknesses and briefly state how a TQM approach would attempt to address each one.

8 Quality improvement approaches

Other formal approaches to quality procedures which use predefined frameworks exist.

8.1 EFQM

The European Foundation for Quality Management's Excellence Model is a total quality model, used as a world-class benchmark for quality management. As a conceptual framework, the Excellence Model recognises that processes are the means by which an organisation harnesses the talents of its people to produce results.

Leadership drives *policy and strategy, people, partnerships and resources* – which in turn shape *processes* – which lead to excellence in *key performance results (people, customer* and *society).* The model attaches criteria and weightings to each of these nine elements, as the basis of organisational self-assessments and third-party awards. The criteria are summarised as follows.

<table>
<tr><td>ENABLER CRITERIA</td><td>What is the approach in this area and is it appropriate?
Does the approach support the organisation's overall aims?
How widely used is the approach?
How is the approach reviewed?
What improvements are undertaken following review?</td></tr>
<tr><td>Leadership
How the behaviour and actions of leaders support a culture of excellence</td><td><ul><li>Leaders develop the mission, vision and values and are role models of a culture of excellence.</li><li>Leaders are personally involved in ensuring the organisation's management system is developed, implemented and continuously improved.</li><li>Leaders are involved with customers, partners and representatives of society (stakeholders).</li><li>Leaders motivate, support and recognise the organisation's people.</li></ul></td></tr>
<tr><td>Policy and strategy
How effectively these are formulated and deployed into plans and action</td><td><ul><li>Policy and strategy are based on the present and future needs and expectations of stakeholders.</li><li>Policy and strategy are based on information from performance measurement, research, learning and creativity-related activities.</li><li>Policy and strategy area developed, reviewed and updated.</li><li>Policy and strategy are deployed through a framework of key processes.</li><li>Policy and strategy are communicated and implemented.</li></ul></td></tr>
<tr><td>People
How the organisation develops and realises the potential of its human resources</td><td><ul><li>People resources are planned, managed and improved.</li><li>People's knowledge and competencies are identified, developed and sustained.</li><li>People are involved and empowered.</li><li>People and the organisation have a dialogue.</li><li>People are rewarded, recognised and cared for.</li></ul></td></tr>
<tr><td>Partnerships and resources
How effectively and efficiently the organisation manages its resources</td><td><ul><li>External partnerships are managed.</li><li>Finances are managed.</li><li>Buildings, equipment and materials are managed.</li><li>Technology is managed.</li><li>Information and knowledge are managed.</li></ul></td></tr>
<tr><td>Processes
How the organisation manages and improves its processes</td><td><ul><li>Processes are systematically designed and managed.</li><li>Processes are improved, as needed, using innovation in order to fully satisfy and generate increasing value for customers and other stakeholders.</li><li>Products and services are designed and developed based on customer needs and expectations.</li><li>Products and services are produced, delivered and serviced.</li><li>Customer relationships are managed and enhanced.</li></ul></td></tr>
</table>

RESULTS CRITERIA	What has performance been over a period of time?
	How does performance compare against internal targets and other organisations?
	Were the results caused by the approaches described in the enabler criteria?
	To what extent do the measures cover the range of the organisation's or business area's activities?
Customer results	• Perception measures: what is the customers' perception of the organisation? • Performance indicators: how good are the drivers of (key factors in bringing about) customer satisfaction?
People results	• Perception measures: what are employees' perceptions of the organisation? • Performance indicators: how good are the drivers of employee satisfaction?
Society results	• Perception measures: how does society and the local community perceive the organisation? • Performance indicators: what results have been achieved relating to community and environmental concerns?
Key performance results	• What is the organisation achieving in relation to its planned performance?

Since 1995, the experiences with the **EFQM** excellence model were initiated by training, the design of quality tools and application guidelines, and actions related to criteria of the **EFQM** model.

Four assessment cycles in which most of the organisations participated were completed. Scores for most of the criteria improved, particularly in 'processes'. The overall patients' satisfaction was higher than 89% in all settings, in most of the cases higher than 95%. Ten organisations (32%) exceeded 400 points in an external evaluation with the EFQM excellence model, and 2 (6%) 500 points. Eighty-three per cent of hospitals have some ISO-certified areas of activity. In the primary care setting, 40% of people were attended in a certified centre.

As a consequence, stimulating actions towards quality have resulted in progressive implementation of the EFQM model, this approach being possibly related to positive evolution of some outcomes. Key factors identified have been pursuing the objective of total quality management during several years and the assignment of the resources for training and implementation of quality systems.

Discussion

Study buddy

Make a very general informal assessment of your organisation using some of the criteria listed in the EFQM table above.

- What immediate areas for improvement (if any) might you highlight for later follow-up?
- What areas strike you as most relevant to, or dependent upon, marketing input and activity?
- Compare and contrast this with your study buddy.
- Discuss with your cohort the value of the EFQM Model.

8.2 Benchmarking

> **Definition**
>
> **Benchmarking** is the establishment, through data gathering, of targets and comparators, where relative levels of performance (and particularly areas of underperformance) can be identified.

Benchmarking essentially involves comparing your operations to somebody else's.

Bendell *et al* (1993) distinguish four types of benchmarking.

- **Internal benchmarking**

    - Comparison with high-performing units or functions in the same organisation.

    - One division's marketing function might be benchmarked against another, for example, to lift standards and consistency within the organisation.

- **Functional benchmarking**

    Comparison with a best-practice external example of a function, regardless of industry. This is also known as operational or process benchmarking. An IT company might benchmark its marketing function against that of Apple Computers or Coca Cola, say, as a way of lifting its processes to best-practice level.

- **Competitor benchmarking**

    Comparison with a high-performing competitor in key areas which impact on performance: for example, Dell comparing its product reliability, time-to-market for innovation or B2B marketing with Hewlett Packard.

- **Generic benchmarking**

    - Comparison of business processes across functional boundaries and industries.

    - The organisation can benchmark itself against 'excellent' companies or an excellence framework (such as the EFQM), or an organisation with a reputation for learning, innovation, relationship marketing – or whatever the organisation is interested in pursuing.

8.2.1 Benefits of benchmarking

Some of the potential benefits claimed for benchmarking include the following.

- It enables the firm to assess its existing performance and position, and particularly its competitive strengths and weaknesses or areas in which it falls short of best practice.

- It focuses on improvement in key results areas (critical success factors) and sets targets designed to be realistic (since other organisations have achieved them) yet challenging (since the benchmarking organisation hasn't yet achieved them).

- It replaces an *ad hoc* or subjective approach to performance measurement, improvement and competitive advantage-seeking with a set of objective, systematic criteria.

- The process is generally carried out by the managers who have to implement and live with the changes: this should help to secure stakeholder buy-in.

- It stimulates research and feedback-seeking about the organisation's strengths, weaknesses and core competences.

- It generates new ideas and insights 'outside the box' of the organisation's usual ways of thinking and acting. Analysis of competitor products and processes, for example, may give rise to ideas which can be learned from or imitated.

It should be noted that there are also limitations and drawbacks to benchmarking as an approach to performance measurement and improvement.

- It may suggest that there is one best way of doing business. However, the organisation needs to take a contingency approach to problem-solving, suitable to its own resources, strengths, weaknesses and environment – rather than an 'off-the-shelf' approach drawn from other organisations' experience.

- The benchmark may be 'yesterday's solution to tomorrow's problem'.

- It is a reactive, 'catch-up' or 'me too' exercise, which may not result in the seeking and development of unique, distinctive and hard-to-imitate core competences.

- It depends on accurate and detailed information about competitors (which may pose research and ethical challenges), and on appropriate analogies/comparisons in other industries (because of the need to compare like with like).

8.2.2 The process of benchmarking

Oakland (1991) puts forward a 15-step benchmarking process.

Plan		
	1	Select the function, unit or process to be benchmarked.
	2	Identify the exemplar of best practice, key competitor or successful partner (using industry analysis, customer feedback or a benchmarking consultant).
	3	Identify the criteria to be benchmarked (based on critical success factors).
	4	Establish a benchmarking project team (clear roles, authority and resources).
	5	Plan methods for data collection (surveys, interviews, documents, visits).
	6	Conduct research.
	7	Manage direct contacts with the target organisation (interviews, visits etc).
Analyse	8	Collate and analyse benchmark data to compare performance: identify performance 'gaps'.
	9	Create a 'competence centre' and knowledge bank: document information.
	10	Analyse underlying cultural, structural and managerial factors that enable performance to benchmark standard (ie not just performance measures).
Develop	11	Develop new performance standards, targets and measures to reflect desired improvements.
	12	Develop systematic action plans (change management programmes, resource plans etc with time-scales, accountabilities and monitoring/review processes).
Improve	13	Implement the action plans.
Review	14	Continuously monitor and/or periodically review progress and results.
	15	Review the benchmark process for future learning.

8.3 Six Sigma

Six Sigma is a disciplined application of statistical problem-solving tools to identify and quantify waste and indicate steps for improvement (Brue, 2005). It uses the define, measure, analyse, improve and control 'DMAIC' model for process improvement.

- **Define** an opportunity or need for improvement: a project team is formed and given the responsibility and resources for solving the problem.

- **Measure** current performance: gather data that describes accurately how the process is currently working and analyse it to produce some initial ideas about what may be causing the problem.

- **Analyse** the opportunity for improvement: generate and test theories as to what might be causing the problem, to identify root causes.

- **Improve** performance: remove root causes by designing and implementing changes to the 'offending' process.

- **Control** performance by designing and implementing new controls to prevent the original problem from returning.

Six Sigma is designed to help an organisation focus on: understanding and management of customer requirements; aligning key business processes to achieve those requirements; utilising rigorous statistical data analysis to minimise variation in those processes; and thereby driving sustainable improvement in business processes.

The method can be outlined as follows.

- Identify and prioritise characteristics of the product or service that are critical to quality (CTQ) for customers.

- Identify and focus on factors and variables in the process that most influence those characteristics ('vital few factors').

- Calculate the variation in the process, when it is operating normally. Six Sigma uses standard deviation (sigma) to measure the variation of values (eg time per unit, or defects per x units produced) from the mean (average).

- A score of 'six sigma' signifies near perfection: 99.999% of items are within specification (3.4 defects per million opportunities).

- Define detailed performance standards and tolerances for the key variables, establishing how much variation is acceptable to the (internal or external) customer of the process. There should be a lower specification limit (LSL) and upper specification limit (USL) within which outputs should fall.

- Set control limits – a lower control limit (LCL) and upper control limit (UCL) which represent the minimum and maximum inherent limits of the process. If these are within the pre-determined limits, the process is capable of meeting specification. If either or both of the control limits are outside the specification limits, the process won't be able to consistently meet specification.

- Implement process control to minimise variations in the vital few factors, aiming for 6 sigma performance. Workers know the expected range of variation: if measurements taken from samples of the process are within the specification limits, the process is in control – and if not, it can immediately be halted and problems rectified.

- Involve management and staff in the process, to create a quality-focused learning culture.

Global case study

The Toughest Stretch Goal

Jack Welch was told that Six Sigma, the quality program pioneered by Motorola, could have a profound effect on GE quality. Although sceptical at first, the GE Chairman initiated a huge campaign – in the GE Way, a way that had never been done before – to infuse quality in every corner of the company. Welch called six sigma the most difficult stretch goal GE had ever undertaken.

Within four years, "we want to be not just better in quality, but a company 10,000 times better than its competitors," he announced. "We want to change the competitive landscape by being not just better than our competitors, but by taking quality to a whole new level. We want to make our quality so special, so valuable to our customers, so important to their success that our products become the only real value choice."

Welch made an official announcement launching the quality initiative at GE's annual gathering of 500 top managers in January 1996. He called the program "the biggest opportunity for growth, increased profitability, and individual employee satisfaction in the history of our company." He set a goal of becoming a six sigma quality company – producing nearly defect-free products, services, and transactions – by the year 2000. The results achieved over the first two years (1996–1998):

- Revenues rose to $100 billion, up 11%
- Earnings increased to $9.3 billion, up 13%
- Earnings per share grew to $2.80, up 14%
- Operating margin rose to a record 16.7%
- Working capital turns rose sharply to 9.2%, up from 1997's record of 7.4

http://www.1000ventures.com/business_guide/cs_quality_six-sigma_ge.html

8.4 The PDCA (Deming) cycle

The Plan-Do-Check-Act or PDCA cycle is a systematic approach to problem-solving, known as the Deming cycle. It is designed as a 'universal improvement methodology': an approach to process improvement – but also to problem-solving and change management in general. It is about learning and on-going improvement: learning what works and what does not in a systematic way, in a continuous cycle.

- **Plan** what is needed

- **Do** it

- **Check** that it works (or what worked and what didn't) by measuring results and comparing them with the plan

- **Act** to correct any problems or improve the process.

The idea is to run a 'pilot project' for the change prior to full implantation: if it doesn't deliver the desired or expected outcomes, you can adjust accordingly – without damage to your credibility or wasted resources. Essentially you plan, test and incorporate feedback before committing to implementation.

8.4.1 Deming: 14 points for improving quality

Deming's (1986) 14 points for quality improvement stressed the need for statistical control methods, participation, education, openness and improvement.

1   Create a constancy of purpose.
2   Adopt a new quality-conscious philosophy.
3   Cease dependence on inspection.
4   Stop awarding business on price.
5   Continuous improvement in the system of production and service.
6   Institute training on the job.
7   Institute leadership.
8   Drive out fear.
9   Break down barriers between departments.
10   Eliminate slogans and exhortations.
11   Eliminate quotas or work standards.
12   Give employees pride in their job.
13   Institute education and a self-improvement programme.
14   Put everyone to work to accomplish it.

9 Service quality

Definition

A **service** may be defined as: an activity or benefit that one party can offer to another that is essentially intangible and does not result in the ownership of anything. Its product may or may not be tied to a physical product.

9.1 Perceived service quality

Chris Grönroos introduced the concept of 'perceived service quality' and developed a widely cited model of service quality (Grönroos, 2000).

Grőnoos (1984) Service Quality Model

The model suggests that the quality of a given service is the outcome of an evaluation process by which consumers compare what they expected to receive with what they perceive that they have actually received.

Their expectations are influenced by factors such as marketing mix activities, external traditions and customs, ideology and word-of-mouth communications – as well as their own previous experience with the service or organisation.

In terms of perceived service quality, Grönroos (2000) suggests that there are two principal components of quality – technical and functional – with a third, image, acting as a mediating influence.

- **Technical quality** is what the customer is left with when the service provision has finished. For example, in education this would be the level of attainment and understanding achieved at the end of the course. This can be fairly easy measured by the consumer.

- **Functional quality** is how the consumer receives the technical quality in interactions with the service provider. Service quality is not just dependent on what is received, but on how it is received. This emphasises the importance of service interactions, customer-facing employees and the management of the service experience – and is harder to measure than technical quality.

Customers' expectations and perceptions of the service are influenced by their perception of the company: its image. If a manager has a positive image of a business school, but then has a negative experience with, say, the Accounts team, the manager may still perceive the service to be satisfactory because (s)he

will find excuses for the problem. Similarly, a negative image may increase perceived problems with
service quality.

9.2 Service gaps – The PZB model

Parasuraman *et al* (1988) developed a widely applied model of service quality, via interviews with 14
executives in four service businesses and 12 customer focus groups. The findings highlighted five
potential quality gaps.

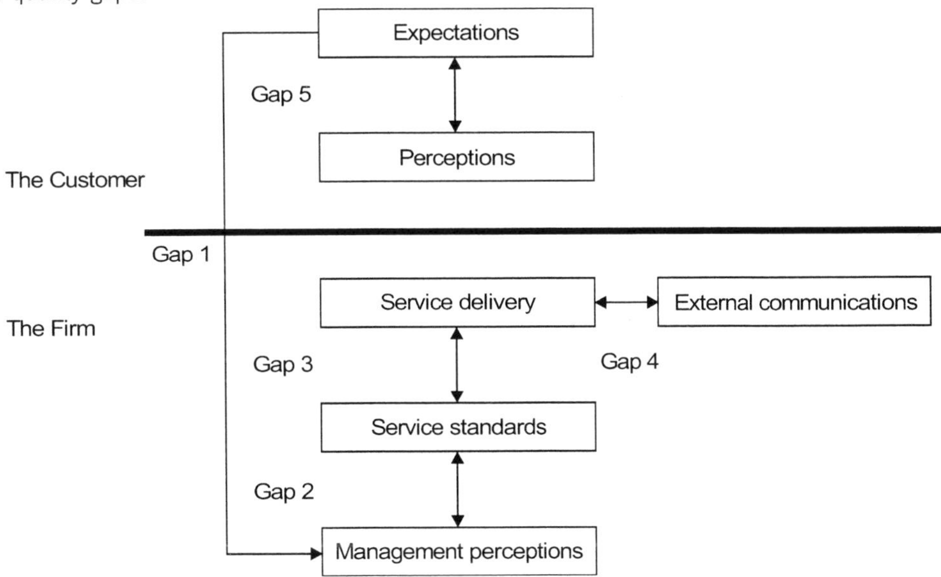

Service gaps

Gap 1 **Consumer expectations v management perceptions**	Managers' definition of value may not be the same as that of customers. They may not know what features connote high quality for customers, what features a service must have or what levels of performance are required by customers. **Action required:** • Market research programmes • Improvements based on customer comment and complaints • Strategies for service recovery • Improvements based on front line staff experience and suggestion
Gap 2 **Management perceptions v service quality specification**	Resource constraints, market conditions and/or management indifference may lead to managers' understanding of required service quality *not* being translated into service level agreements or quality specifications. **Action required:** • New concepts of service rather than merely improving old ones • Attention to physical evidence of quality • Customer-focused activity goals
Gap 3 **Service quality specifications v service delivery**	Service level agreements and quality specifications may not translate into actual service levels. Employees may not be willing or able to perform to the specified standard, work may be poorly organised, supervised or resourced. **Action required:** • Define job roles and priorities clearly • Provide proper training • Build teams and team working • Empower frontline service staff • Improve technology • Recruit, train and reward for policy improvements

Gap 4 Service delivery v external communications	External communications may create exaggerated or uninformed expectations of service quality – setting customers up for disappointment. **Action required**: • Improve communications between internal staff • Educate customers as to what to expect from the service • Develop service rules – but do not over-promote to customers • Use marketing communications to emphasise what is actually delivered
Gap 5 Expected service v perceived service	What customers perceive they have received may fall short of what they expected: ie there is a shortfall in perceived service quality. **Action required**: • Research customer expectations • Measure service performance against expectations • Specify service quality/levels according to expectations • Manage and continuously improve service levels to match expectations

9.2.1 Dimensions of service quality

Parasuraman *et al* (1998) developed the SERVQUAL questionnaire, which claims to be a global measure of service quality across all service organisations. It measures five generic criteria that consumers use in evaluating service quality.

- **Tangibles**: physical facilities, equipment, appearance of personnel
- **Reliability**: ability to perform the promised service dependably and accurately
- **Responsiveness**: willingness to help customers and provide prompt service
- **Assurance**: knowledge and courtesy of employees and their ability to convey trust and confidence
- **Empathy**: caring, individualised attention

Respondents are asked first to give their expectations of the service on a seven-point scale, then to give their evaluation of the actual service on the same scale. Service quality is then calculated as the difference between perception and expectation, weighted for the importance of each item.

Once an organisation knows how it is performing on each of the dimensions of service quality, it can use a number of methods to improve its performance, as necessary.

Activity 6

Make a very general informal assessment of your organisation using the PZB model.

- What areas strike you as having the greatest gaps?
- What immediate areas for improvement (if any) might you highlight for later follow-up?

There are many dimensions in service delivery, before and after the service encounter itself.

- The creation of a corporate culture which expresses and models customer-focused values, and reinforces those values through its selection, appraisal and reward systems, and the messages it sends employees at every level.

- The creation of service-supporting internal relationships and internal marketing: the recruitment of skills customer-facing people; the supply of appropriate training; the empowerment of staff to take decisions that will satisfy and retain customers; and the reward and recognition of staff who deliver outstanding service.

- Gathering, analysing, communicating and acting on customer feedback. Feedback and adjustment (addressing customer concerns and complaints) are crucial in minimising dissatisfaction and demonstrating commitment to customer value.

- Constructive handling of problems and complaints (sometimes called 'service recovery') may lead to restored satisfaction – and even strengthened relationships, because of the provider's demonstrated commitment.

- Establishing a partnership approach to relationships with customers, suppliers and intermediaries (distributors, retail outlets, call centres) in order to support high levels of service at all links in the value-delivery chain.

- Ensuring promise fulfilment: Making realistic promises (to manage customer expectations); enabling staff and service systems to deliver on promises made; and keeping promises during service encounters.

- Offering support services (eg warranties, servicing, user training and help-lines) to facilitate customers in using the product safely and satisfyingly, and support them through changes and difficulties.

- Establishing customer-friendly systems. It is no good expecting staff to give great service to customers if the systems, procedures, technology and information flows do not support their efforts.

1 Evaluate the importance of organisational structures in delivering marketing value, focus and creativity that aligns the organisation to the vision

- An organisation is the sum total of the ways in which it divides its labour into distinct tasks and then achieves co-ordination among them.

- Henry Mintzberg (1984) provided a framework and language for discussing organisation structure, by categorising the building blocks of organisation.

- Organisational structure is shaped by many factors. There are certain internal principles and dynamics which determine some elements of structure.

- A number of contingent variables influence structural choices and organisational development.

- In considering organisational structures and designs it is also important to recognise that organisations are required to adapt, reflect and change as they grow and evolve.

- Organisation charts act as a guide to explain how different task roles and positions relate to each other, and how they are co-ordinated and integrated.

2 Recommend how an organisation could be structured to deliver competitive advantage and organisational success

- In most organisations, tasks and people are grouped together in some rational way: on the basis of specialisation or customer base.

- Some organisations group activities on the basis of products or product groups or 'lines'.

- In a variant on the product-based approach, managers may take responsibility for particular markets or customer segments

- Geographical structures are ideal if markets vary considerably and beneficial where products are largely homogenous and there is a requirement for fast and efficient distribution.

- A matrix structure can consist of two organisational structures (eg geographic and product structures) intersecting each other with dual reporting relationships.

- A major concern of management when considering organisational structure is division of labour or specialisation and the degree of centralisation and decentralisation.

- Some recent trends have emerged from a focus on responsiveness and flexibility as key organisational values.

- The flattening of hierarchies does away with levels of organisation which lengthened lines of communication and decision-making and encouraged ever-increasing specialisation.

- Charles Handy's 'shamrock organisation' is gaining ground as a workable model for a leaner and more flexible workforce, within a controlled framework (Handy, 1989).

- Boundaryless structures are being developed where horizontal (status) barriers are removed or softened by delayering and participative decision-making, and vertical (functional) barriers by empowered cross-functional team working – in order to align business processes.

3 Evaluate the issues that need to be taken into consideration in designing an organisation structure that aligns itself to the vision

The choice of organisational design and structure will depend partly on the preferences of top management, and the style it likes to adopt.

Key processes must also be integrated together with the organisational structure

4 Assess the nature of quality

- Quality is a measure of the features and characteristics of a product or service which affects their ability to meet stated or implied needs.

- Quality means different things to different operations.

- Marketers need to identify 'quality gaps': shortfalls between what customers expect or perceive, and what the organisation provides.

- A number of writers have been influential on quality thinking in recent decades; Juran, Deming, Ishikawa, Taguchi, Crosby etc.

5 Evaluate quality systems and processes

- The business case for quality management includes lower costs, maintenance of competitive advantage, enhanced brand positioning and customer retention.

- Quality management is concerned with controlling activities with the aim of ensuring that products or services are fit for purpose, and meet specifications.

- Quality management has come to represent not just a managerial function, but a managerial orientation – affecting all areas of the organisation's structures and processes.

- The key embodiment of this orientation is Total Quality Management (TQM).

- A quality management system is a set of co-ordinated activities which will direct and control an organisation in order to continually improve the effectiveness and efficiency of its processes and performance.

6 Understand the concept of Total Quality Management

- Total Quality Management is an orientation to quality in which quality values and aspirations are applied to the management of all resources and relationships.

- There are seven key principles and values of TQM.

- Quality management involves the on-going, continual examination and improvement of existing processes: 'getting it more right, next time'.

- A quality circle is a team of workers from different levels and functions of an organisation which meets regularly to discuss issues relating to quality, and to recommend improvements.

- TQM is susceptible to various adverse perceptions.

7 Explore quality improvement approaches

- The European Foundation for Quality Management's Excellence Model is a total quality model, used as a world-class benchmark for quality management.

- Benchmarking is the establishment, through data gathering, of targets and comparators, where relative levels of performance (and particularly areas of underperformance) can be identified.

- Six Sigma is a disciplined application of statistical problem-solving tools to identify and quantify waste and indicate steps for improvement.

- The Plan-Do-Check-Act or PDCA cycle is a systematic approach to problem-solving.

8 Critically evaluate the role of service quality

- Grönroos (2000) introduced the concept of 'perceived service quality' and developed a widely cited model of service quality.

- Customers' expectations and perceptions of the service are influenced by their perception of the company: its image.

- Parasuraman *et al* (1990) developed a widely applied model of service quality highlighting five potential quality gaps.

Activity 1

Greiner's organisational growth model (1972) provides a framework against which managers can derive a framework within which they can understand their circumstances as a basis for decision-making. He identifies five stages of growth. The answer will depend on your organisation.

Activity 2

A matrix structure can consist of two organisational structures (eg geographic and product structures) intersecting each other with dual reporting relationships.

Members retain a working relationship within their individual disciplines at the same time as reporting to a project leader for the specific project in hand. The answer will depend on your organisation.

Activity 3

A major concern of management when considering organisational structure is division of labour or specialisation (ie who does what, where does responsibility lie, what authority/control over resources do individuals have). This answer will depend on the context of your own organisation.

Activity 4

The answer will depend on your organisation.

Activity 5

There is a tendency, when something doesn't work properly, to try doing it again. However, if something hasn't worked properly in the past, there is no reason to suppose that it will work in the future. TQM encourages an attitude that seeks to change how things are done.

Understanding competitive positioning is largely a strategic issue for management, but without a proper understanding of competitive positioning, and the need to have a competitive position in their markets, companies will not survive – particularly in weak or declining industries. A TQM approach includes working out ways of achieving customer satisfaction in order to compete effectively.

In many companies, each department or function in an organisation thinks only about its own needs and doesn't think about the needs and expectations of other departments (that might be its 'internal customers' or 'internal suppliers'). This can create problems for the other department to resolve.

Management has a tendency to treat people like robots and employees who are treated in this way will tend to act accordingly – with little concern for the work they are doing.

Gold-plated taps in a bath represent a high grade of materials, but if the taps have faulty washers and leak, the quality will be poor. High grade does not ensure good quality. Equally, a low grade product can provide better quality than a high-grade item. For example, uniforms worn by workers might be of a better quality ('fit for the purpose') if they are made from a cheaper, hard-wearing material, than if they are made from more expensive material that is more easily torn or more difficult to clean.

Having an acceptable level of quality implies that some errors are tolerable. A TQM approach is that everything should be done right the first time.

Managers often enjoy dealing with problems (fire-fighting) as it gives them the sense of being 'in charge'. A TQM approach is that managers waste their time if they have to deal with problems that should not have arisen in the first place.

There is a tendency for individuals to be unconcerned about errors they make, so long as the problem does not affect them, but someone else. With a TQM approach, employees are encouraged to think of other departments as internal customers, and accept responsibility for poor service to their customers.

Activity 6

This will depend upon the research you undertake on your own organisation.

Brue, G., (2005). *Six Sigma for Managers*. Maidenhead: McGraw Hill Professional Education.

Burns, T. & Stalker, G. M., (1961). *The Management of Innovation*. London: Tavistock.

Child, J., (1972). 'Organisational structure, environment and performance: The role of strategic choice'. *Sociology*, Vol. 6, pp. 1-22.

Crosby, P. B., (1979). *Quality is Free*. London: Mentor.

Deming, W. E., (1986). *Out of the Crisis*. Boston: Massachusetts Institute of Technology.

Drucker, P., (1955). *The Practice of Management*. London: Heinemann.

Garvin, D. A., (1987). 'Competing in eight dimensions of quality' in *Harvard Business Review*, November – December.

Greiner, L., (1972). 'Evolution and revolution as organizations grow'. *Harvard Business Review*, July, pp. 37-46.

Grönroos, C., (2000). *Service Management & Marketing,* 2nd edition. London: John Wiley.

Handy, C., (1989). *The Age of Unreason*. Harmondsworth: Penguin.

Huczynski, A. & Buchanan, D., (2001). *Organizational Behaviour: An Introductory Text,* 4th edition. Harlow: FT Prentice Hall.

Ishikawa, K., (1991). *What is Total Quality Control? The Japanese Way*. Harlow: Prentice Hall.

Juran, J.M., (1998). *Juran on Planning for Quality*. New York: Free Press.

Kotler, P., (2002). *Marketing Management,* 11th edition. Harlow: FT Prentice Hall.

Milkovich, G.T. and Brodie, J.W., (1997). *Personnel/Human Resource Management: A Diagnostic Approach,* 8th edition. Homewood IL: Irwin.

Mintzberg, H., (1984). *The Structuring of Organisations*. Englewood: Prentice Hall.

Mullins, L., (1999). *Management & Organisational Behaviour,* 5th edition. Harlow: FT Prentice Hall.

Oakland, J. S., (1991). *Total Quality Management: The Route to Improving Performance*. Oxford: Butterworth-Heinemann.

Parasuraman, A., Zeithaml, V. A. & Berry, L., (1988). 'SERVQUAL: a multiple-item scale for measuring customer perceptions of service quality' in *Journal of Retailing*. Spring, pp.12–40.

Pedler, J., (1991). 'From learning organisation to knowledge entrepreneur'. *Journal of Knowledge Management*. Vol. 6, p.258.

Peters, T., (1989). *Thriving on Chaos*. London: Macmillan.

Reichheld, F. F., (1996). *The Loyalty Effect*. Boston: Harvard Business School.

Sheth, J., (2007). *The Self Destructive Habits of Good Companies*. Pennsylvania: Wharton School Publishing.

Taguchi, R., (1986). *Introduction to Quality Engineering*. Tokyo: Asian Productivity Organisation.

Zeithaml, V., Parasuraman, A. & Berry, L., (1990). *Delivering quality services*. New York: Free Press.

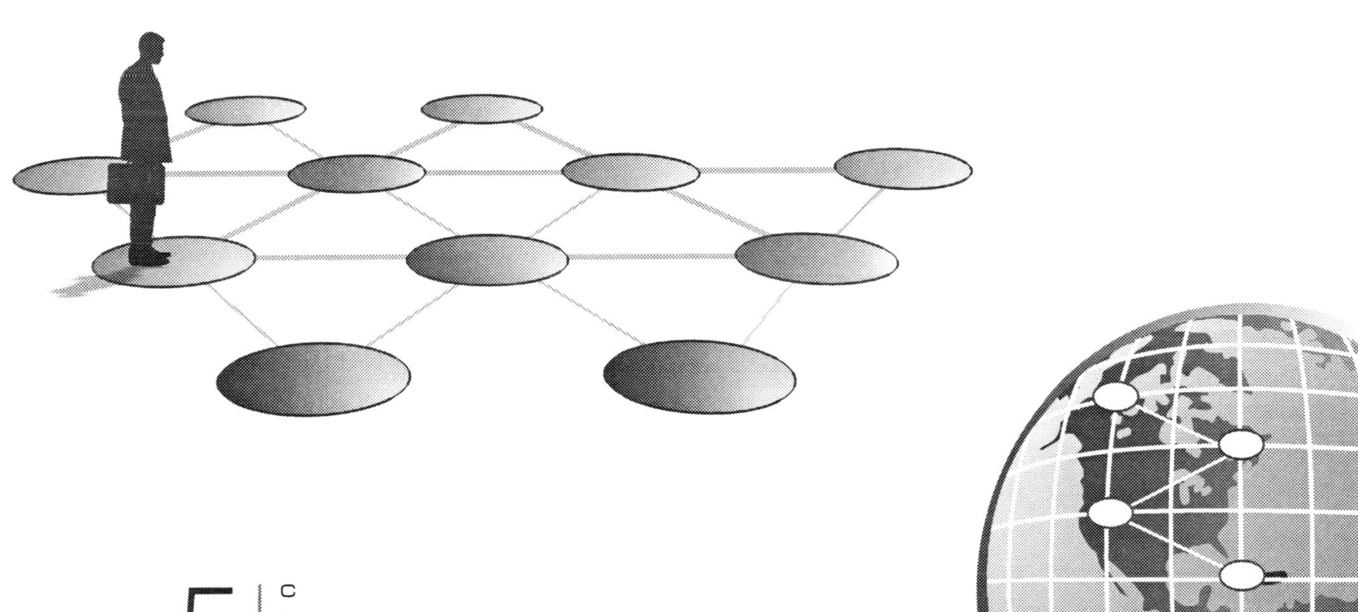

5 chapter

Managing, implementing and monitoring change

Change for most organisations is now a fact of life although the reasons might vary. Change is usually met with apprehension or 'oh no, not again' and neither of these responses are likely to lead to employees enthusiastically embracing change. The change itself is not really the problem, particularly today in many cultures where change is common, the problem is how change is announced and managed. Typically the 'announcement ' starts with rumours which begins a negative spiral before anything is officially said and managing the truth and the gossip is left to chance, to be fitted in at the manager's discretion. Change is not without pain but it can be managed and implemented with integrity. In this chapter we review what internal and external factors trigger change, how we can develop a process for effective change and how we need to plan for change.

We then evaluate the role and importance of a clear organisational vision in the context of change management. Next, we consider the stages of transition that an organisation and individuals go through and the techniques that need to be deployed in order to mange the transition successfully.

We finally review the importance of monitoring change and the mechanisms that can be deployed to evaluate the progress and success of the change process.

Contents

Chapter learning outcomes

In this chapter you will cover the following:

- Evaluate the external and internal drivers of change and barriers to change
- Critically assess methods of change and its impact on the marketing strategy
- Rationalise a process for managing change
- Evaluate how to best prepare a plan for change

1 Drivers of change

The need for change can be triggered by either internal or external triggers.

1.1 External triggers of change

External changes occur in seven broad areas all of which are linked either tentatively or strongly. These ripples in a pool mean a change in one area may have an immediate and direct influence over a change in another or the change or influence may be almost imperceptible and oblique.

Any one of the following factors, at different times, will lead in changing events. Marketing is responsible for identifying trends and the potential for change.

Customers	**Basic needs** – how will these change, what will change them, with what are they disenchanted, what are their expectations etc?
	Demographics – what changes are taking place, what groupings can we expect, what are the trends etc?
	Life stages – what are they (characteristics), how are they changing etc?
	Habits and usage – where and when are products used, why, how?
	Attitudes – towards industry, environment, society.
Politics and the law	**Policy** – for example, a change in public spending.
	Legislation – for example new employment laws.
Economy	**Boom** – employment high, disposable income high, spending high.
	Recession – employment, disposable income and spending low.
Environment	**Climate** – global warming and changes in customer behaviour
	Natural disasters – increase in flooding or drought
Channels	**Current channels** – which are evolving, working well, which are not?
	Future channels – what developments suggest new opportunities, how easy will these be to exploit/implement?
Technology	**Technology as a product/service** – what products and services are/can be used to exploit opportunities and meet needs, what advances are or are likely to be made?
	Technology as a means of delivery – what developments might change the way in which our products or services are delivered, how will this affect the business?
	Technology as a means of production – both in terms of manufacturing capability and business operations? What developments might change the way things are done?

Regulation	**Legislation** – political agendas, lobbying groups (industry, pressure groups). **Deregulation** – eg markets, business practices. **Privatisation** – changing society in terms of State support. **Voluntary codes** – industry and professional institutions, associations establishing codes of practice designed to avoid legislation and over policing and protect customers.
Finance	**Costs** – what are they, will they continue to escalate (in spite of technology), what will be the impact of inflation, taxation? **Pricing** – what role will price play in the future, will pricing wars escalate? **Profit** – where will profits be made in the future (opportunities), how will profits be protected (business operations)? **Funding** – increasing influence of financial markets, likely convergence of financial markets.
Competition	**Performance** – extent of strategic activities and effectiveness of marketing programmes. **Nature of** – likelihood of new entrants and their affect on the industry/markets.
Innovation	**Who are the innovators** – are there particular industries driving innovation, particular professions, people? **Where are the innovations** – is it mainly technology and in what industries, geographic areas, institutions (eg universities, business schools)? **What are the innovations** – is it likely to be the products and services sold, business processes, is it in the way we think and behave?

1.2 Internal triggers of change

We can use the McKinsey 7S Model (Waterman *et al*, 1980) to identify what might trigger change internally.

Shared values	For example, if we want to change to a marketing orientation we might want to change values and behaviour.
Strategy	A change in strategic direction can require a change in, eg competitive position.
Structure	A marketing orientation putting customers at the centre of the business might require a change in structure.
Systems	Changing to a marketing orientation means changing from a product delivery system to a value delivery system.
Style	Leadership and management styles may need to change to reflect different values.
Staff	Learning and development will be important to enable staff to adopt new ways of thinking and behaviour to focus on customer value.
Skills	Training will be required to equip employees with new or different skills, particularly focused on the customer experience.

As with external drivers, internal drivers are linked and a change in one areas affects what is happening in other areas. Leaders need to understand these connections and while they devolve responsibility for managing and implementing change to management, leaders must co-ordinate and integrate what is going on across the organisation.

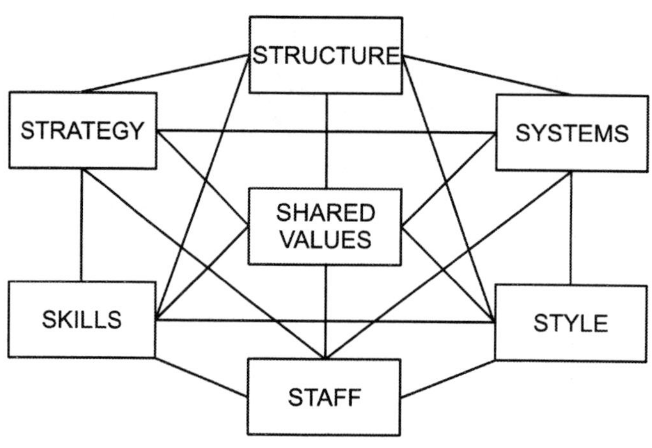

McKinsey 7S model

Study buddy

Thinking about your own organisation undertake an audit using the 7S model and consider the strengths and weaknesses your organisation has for each element. Compare and contrast with your study buddy and share with the discussion forum the value of the 7S Framework.

Identify external and internal drivers of change for your organisation.

1.3 Identifying barriers to organisation change

As we have acknowledged reactions to change are varied and can be positive particularly if the organisation is clearly in difficulty. However, negative reactions are the result of years of poor change management **and** of unnecessary change.

There is no question that in some industries continuous change is essential, for example the technology industry. However, that is not true of all industries and organisations. The unnecessary change seems to be a result of a knee jerk reaction to the external environment. The marketplace is inevitably always changing and the skill of leaders is to recognise when it is appropriate to change the organisation in response to external changes and when it is not.

Continuous improvement is desirable and it could be argued this leads to continuous change. Continuous improvement is often small and very localised and may not require change beyond a person undertaking a task. There may be a great deal of these across the organisation but they do not usually cause negative reactions or meet with resistance unless there is a problem with the employee, the manager or between them both. When the scope of improvements is larger and widespread this will lead to more momentous change.

Among barriers to change are:

- Legacy of previous change programmes if they were poorly managed and did not lead to obvious improvements.

- Existing leadership and management styles and relations with employees. If these are not healthy change will be difficult.

- Leadership and management skills if they have no experience of change or have handled change badly.

De Wit and Meyer (2005) describe leadership activities in broadly three ways:

1   **Strategic thinking**: conceptualising to develop strategic direction
2   **Strategy formation**: creating the right environment for strategy development
3   **Strategic change**: what needs to be done to change the organisation to deliver the strategy

Leadership is naturally focused on the future, developing a vision for that future and anticipating and scoping strategies to deliver successful outcomes. It can be difficult at this level to understand the practicalities of change but leaders should not be preoccupied with this as they work in collaboration with management who work with employees to design and implement change successfully. Leaders and managers have different roles but together change the organisation.

If it is essential to continually change, managers cannot assume it is not stressful and de-motivating and have a duty to reduce stress. Motivation achievement is a typical motivator so at some point there needs to be a sense of achieving something, arriving somewhere or continuous change can become like a treadmill.

- It is important therefore that managers check the validity of 'continuous'.
- Is it relevant to us?
- Is it a priority?
- How much needs to change if it is relevant and is a priority?

In addition managers must provide a sense of achievement and arrival by:

- Acknowledging completion, an end to a particular project.
- Recognising and rewarding success.
- Learning from the experience.

Discussion

Study buddy

Discuss with your Study buddy they key changes that have taken place in each of your organisations in recent years and why? Can you identify the purpose or objective of this change? Evaluate the success of the changes that have taken place.

2 Change and the impact on marketing strategy

The implementation of a marketing strategy will attempt to bring about strategic change. Sometimes in order to respond to environmental change and maintain a competitive advantage a number of strategic changes need to be executed in a variety of areas to keep the organisation aligned with market demands.

Where this occurs this may be referred to as strategic renewal (De Wit and Meyer, 2005). This process might consist of a few large steps or numerous small ones. The size of the change is referred to as the magnitude of change and can be divided into two component parts:

1   **Scope of change**

- This can vary from broad to narrow. Change is broad when many aspects and parts of the firm are altered at the same time.

- In most extreme cases this will result in entire organisational processes, culture and people being changed in unison.

- Change can also be much more narrowly focused on a specific organisational aspect or department.

- If many changes are narrowly targeted the total result will be a more piecemeal change process.

2 **Amplitude of organisational changes**

- This can vary from high to low.
- It is high when the change deployed is a radical departure from what went before.
- It is low when the step proposed is a moderate adjustment.

2.1 The pace of change

Strategic renewal takes time. Strategic change measures can be evenly spread out over an extended period, allowing the organisation to follow a relatively steady pace of strategic renewal. It is also possible to cluster all changes into a few short irregular bursts, giving the renewal process an unsteady, stop-and-go pace.

The pace of organisational change can be split into two related parts.

1 **Timing of change**

- Pace of change depends on the moment at which change is initiated.

- Change can vary from intermittent to constant.

- When change is intermittent it is important to determine the right moment to launch a new initiative.

- The need to wait for the 'right time' is often a reason for spreading change activities unevenly over time.

- On the other hand, change can be constant so that the exact moment for implementing a new measure such as the marketing strategy is less important.

2 **Speed of change**

- Pace of change also depends on the time span within which change takes place.

- Speed of change can vary from high to low.

- Where a major change needs to be implemented within a short period of time the speed of change must be high.

- A short burst of action can bring about the intended changes.

- Alternatively, where the change measures are less formidable and the time span for implementation is longer, the speed of change can be lower.

Organisations, therefore, have many different ways of bringing about strategic change and this brings about the question of whether to achieve revolutionary (disruptive) or evolutionary (gradual) change.

2.2 The demand for revolutionary change

Revolution is a process whereby an abrupt and radical change takes place within a short period of time. Such processes do not build on the *status quo* – they overthrow it. Such an approach may be required when 'organisational rigidity' is so deep-rooted that smaller pushes do not bring about the required movement.

Typical sources of rigidity are explored in the following table:

Source	Comment
Psychological resistance to change	• Many people resist change because of the uncertainties associated with it. • It can be necessary to break through this by imposing a new business system on people.
Cultural resistance to change	• People can become immune to the signals that they are out-of-touch. • Can break through this by exposing the organisation to a shocking crisis or imposing a new organisational system.
Political resistance to change	• Each organisational change invariably results in winners and losers. • Can break through this resistance through a management restructure.
Investment lock-in	• Investment in a new product. Technology or activity can result in fixing the organisation to this even if it is not working. • Can be necessary to break through this 'lock-in' by radically restructuring or disposing of the investment.
Competence lock-in	• When an organisation builds a competitive advantage based on a particular set of competences it is natural for that organisation to favour external opportunities based on those competences. • This can lead to an organisation becoming out-dated. • Changing the core competence of the organisation in a comprehensive and radical manner may be the only way to migrate from one set of competences to another.
System lock-in	• Organisations can become locked-in to a particular standard or system. • Lock-in can only be overcome by a 'big bang' transition to another system.
Stakeholder lock-in	• Restrictive long-term commitments can be made to stakeholders (ie long-term contracts with suppliers, long-term warranties with buyers etc). • It might be necessary to aim for a radical restructuring of the organisation's relationships.

Besides the use of revolutionary change to overcome rigidity there may be other triggers that are necessary for revolutionary change.

- **Competitive pressure**

  When an organisation is under intense pressure and its market position is eroding quickly a rapid and dramatic response might be the only response possible.

- **Regulatory pressure**

  Organisations can be put under pressure by government or regulatory agencies to push through major changes within a short period of time.

- **First mover advantage**

  A more proactive reason to implement a revolutionary change is to be the first to introduce a new product, service or technology and to build up barriers to entry for late movers.

Most leaders recognise that their organisations are prone to inertia and acknowledge that it is often necessary to move quickly in response to external pressures or to take advantage of first mover advantage. Most would like their organisation to have the ability to successfully implement revolutionary strategic changes.

Motorola

Over the decades, Motorola's steady evolution has allowed it to meet the challenges of revolutionary changes in business. Change has become part of Motorola's legacy. We call it renewal, a concept that was introduced by our founder, Paul Galvin.

He believed that in order to survive the company had to renew itself continually. Soon after the company's founding in 1928, Motorola phased out of the declining battery eliminator business and ventured into the fledgling market for automobile radios. By the mid-1940s, the car radio was used as a technology platform for entering the two-way radio business. From transistors, Motorola launched into the brand new semiconductor business in 1949. In the early 1970s, the corporation made a strategic decision to concentrate on opportunities in industrial markets. But, by the 1980s Motorola was back in consumer markets as a major supplier of cellular telephones and pagers. Today, with 1992 net sales of $13.3 billion, Motorola is a major player in a fierce electronics revolution of global dimensions. While our product lines have grown, evolved and multiplied, the people values on which Motorola was founded have remained constant. Chief among these is a constant respect for people. Paul Galvin understood that his goals could only be achieved if he attracted people to his vision, people who would share his vision. Motorola's people – "Motorolans" – have made it possible for the company to effectively and repeatedly renew itself over the decades.

http://www.allbusiness.com

2.3 The demand for evolutionary change processes

Evolutionary change processes take the current state as a starting point and constantly modify aspects through extension and adaptation. A new system or strategy can evolve out of the old, as if an organisation is 'shedding its skin'.

This approach is more important where the strategic renewal process hinges on widespread organisational learning. Learning is a relatively slow process and time will be needed to experiment, reflect, discuss, test and internalise the changes.

A more evolutionary approach may be the only option when no one person has enough sway to push through radical changes. To some extent leaders recognise that their organisations need to continuously learn and adapt and acknowledge that they do not have enough absolute power to impose revolutionary change.

Once leaders choose to adopt revolutionary or evolutionary change they will find it very difficult to simultaneously or subsequently use the other.

2.4 Discontinuous v continuous renewal perspective

Some commentators feel that revolutionary change is seen to be at the heart of renewal while evolutionary change can only play a supporting role. This is often referred to as the 'discontinuous renewal perspective'.

Others feel that real strategic renewal must grow out of the existing organisation with a steady stream of adjustments and that, while difficult to sustain, evolutionary change is at the heart of renewal while revolutionary changes are a fall-back alternative. This point of view is referred to as the 'continuous renewal perspective'.

2.4.1 The discontinuous renewal perspective

Proponents of this perspective argue that people and organisations exhibit a natural reluctance to change. Once general policy is determined organisations are inclined to settle into a fixed way of working. This is further cemented after a sustained period of success.

Advocates therefore argue that long periods of stability are necessary for the proper functioning of firms. However, this leads to rigidity. The way to overcome this is to implement co-ordinated, radical and comprehensive measures.

Some proponents of this view argue that episodes of revolutionary change are generally not chosen freely but are triggered by crises (eg introduction of a new technology, novel market entrant etc). However, often a misalignment between the organisation and its environment grows over a longer period of time causing a mounting sense of crisis. As tension increases people become more receptive to submitting to the changes that are necessary.

As long as there is pressure revolutionary change is possible. Leaders will therefore induce a crisis to create the sense of urgency and determination that may be required to get people to change.

Others argue that revolutionary change can be proactively pursued to gain a competitive advantage such as using a break through technology, creating a new business model and so on. Old methods must be discarded before new methods, strategies and so on, can be adopted.

This process is not orderly or protracted but can be disruptive and intense. Therefore organisations must learn to master the skill of ongoing revolutionary change.

So, for example, rapid implementation of new strategy is an essential organisational capability. Such change management requires strong, consistent leadership to rapidly push through any stress that may be related to it.

Battling against rigidity and turning crisis into opportunity become key qualities required by leaders. Those responsible for developing and implementing strategic change need to be aware of the timing of such events.

2.4.2 The continuous renewal perspective

Proponents of this view argue that if organisations 'shift by earthquake' then it is usually their own fault and that such change creates a 'boom-and-bust' cycle. Those organisations that maintain a steady stream of change will win in the end.

Revolutionary change causes unnecessary disruption and dysfunctional crises and only has short-term impact. Continuous renewal has a more long-term orientation and is constantly maintained over a longer period of time.

Three organisational characteristics are important for keeping up a steady pace of change:

- Employees within the organisation are committed to continuously improve
- Everyone in the organisation is motivated to continuously learn
- Everyone in the organisation must be motivated to continuously adapt

Leaders should strive to create flexible structures and processes, an open and tolerant culture and provide sufficient job and career security for employees to accept other forms of ambiguity and uncertainty.

By pursing this approach everyone in the organisation is involved and leaders recognise that although change can be led from the top it cannot simply be Imposed.

Compare and contrast evolutionary v revolutionary change for your organisation, the approach that you think it should take and why.

3 Designing a process for change

Definition

Change management is a deliberate strategy to minimise the negative impact of change thereby reducing resistance and maximising participation to achieve successful design and implementation of change.

So how do you make the transformation from whatever orientation currently exists to a marketing orientation? A process for change management can help keep managers on track and help employees understand the journey.

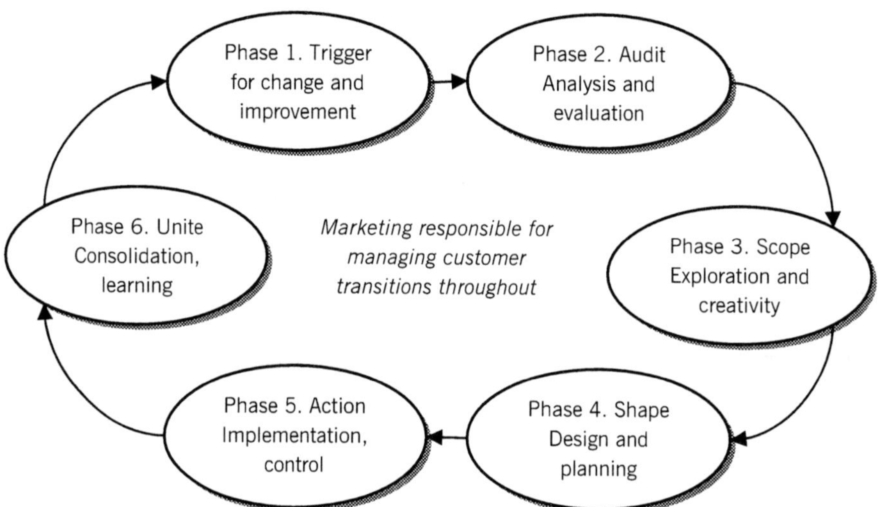

Managing change journey

The challenge for the leader is to pull together the process of managing change with managing people through change. It is easy to focus on operations and tasks during change as there is a sense of urgency about getting things right.

However, we can only get things right if we manage people. Crucially, for the leader, managing the customer transitions throughout change is essential. Hopefully, the customers will not be aware most of the time unless the change is important to them. Where change may impact on them marketing managers need to develop a strategy for managing relationships.

Let us now look at the detailed phases in the overall change process.

3.1 Phase 1 – Trigger for change

We have already considered triggers for change earlier in the chapter.

3.2 Phase 2 – Audit: analysis and evaluation

3.2.1 Announcement (employee concerns and support)

Employees should not become aware of plans for change via the grapevine, gossip and rumour. They should also not become aware of the intention to change after decisions have been made.

Once the need for change has been established there should be a timely announcement as part of a planned communication that reduces the potential for shock.

The first stage of communications is designed to present the requirement for change and invite employees to become involved in the process.

Communications during Phase 2 are designed to help understand the need for change, create dissatisfaction with the current set-up and a desire for, or at least understanding of, change. The task is to interpret new cultural values and the purpose and mission as something meaningful and personal to all individuals at all levels.

The focus should be on:

- Benefits from the employees' perspective, eg new skills, development and their role and the value of their contribution.

- Their association with the organisation's image and reputation.

- Providing support by encouraging discussion, feedback and airing of concerns.

3.2.2 Analysis and evaluation

External analysis focuses on customers, who they are; what they buy; where; how; when they buy and why they buy?

This analysis includes customer behaviour and their motives and needs, however, need to be aware of other environmental factors, for example the nature of industry and competitive situation. Our goal here is to make sure we have a comprehensive understanding of our customers so we understand how we need to respond internally.

Internal analysis focuses on what we do, the purpose of the organisation and its strategic goals. We also review how we do things, operations, policies, procedures, systems and processes and to what extent they meet customer needs or are designed for internal convenience.

This includes the organisation environment and the physical conditions within which employees work. We examine when we do things, timing of activities and events, are they timed for our convenience or the customers?

Finally, we examine why we do things and this will reveal our cultural norms and values. Employees are required to gather this information and their knowledge will be invaluable in establishing the current orientation strengths and weaknesses.

Leaders of change need to work on creating an atmosphere of honesty and openness. If they do not then weaknesses may be ignored or covered up and become an obstacle to achieving a marketing orientation.

The audit should reveal the combination of internal and external drivers of change and restraining forces.

3.2.3 Force field analysis

Lewin (1951) explained that change occurs when change drivers collectively overcome restraining forces. Managers can identify possible drivers and restraining forces and apply weighting to provide an assessment of the change task and likelihood of success or failure. This process encourages managers to think through the issues and problems and plan the change accordingly.

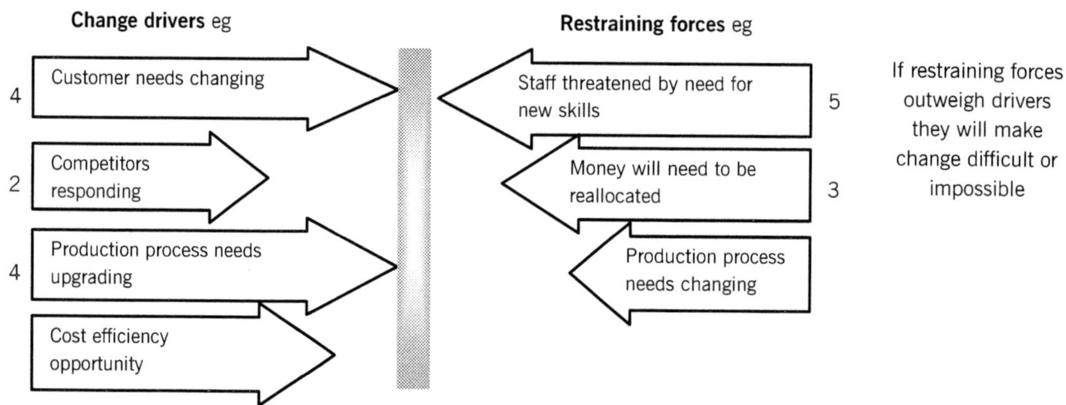

Force field analysis

3.2.4 Leadership, management and cultural differences

Good practices do not just focus on employees. Leadership and management respond to change in different ways. If change is taking place across cultures this will affect the approach to change and way in which it will be managed.

Lasserre and Schutte (2006) suggest that unless we understand the differences in thinking, behaviour and contextual differences we can make little progress.

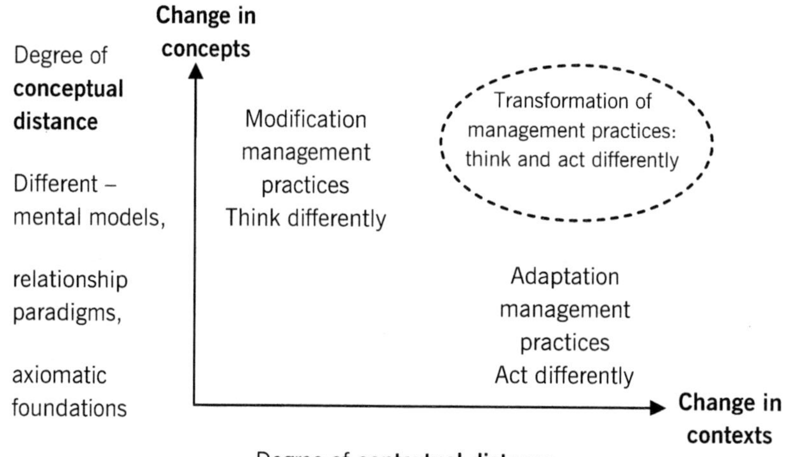

Transforming management practices in different cultures

The goal is to get leaders, managers and employees to think and act differently and managing change across cultures adds another dimension to change management. If the culture is risk averse change will be difficult; if the emphasis is on past glories rather than the future the task is to relate the future to the past. If culture dictates employees defer to managers and do not expect to be consulted on change, adjustments will need to be made to change management.

3.2.5 Understanding employees

Change means different things for different levels of the organisation. Leaders are often the initiators of change (but not exclusively) to pursue their vision and future strategic direction and are rarely interested in the detail of implementation.

Managers are responsible for managing change and will either be campaigners or opponents. Either way they are the mediators between initiators and those who implement change. Employees are the implementers of change and will either be campaigners or opponents.

We need to understand the effect change will have on people including uncertainty; a lack of control over personal performance and increasing frustration; a possible loss of responsibility; possible loss of social networks; lack of security; a sense of abandonment and feeling incompetent as the realisation of newly defined roles emerge.

When change starts employees embark on a journey that runs parallel to daily work routines. It is unsettling if this journey is not understood and people are not properly prepared or managed. It becomes threatening as the journey increasingly impacts on the daily work routines and the social climate.

The personal journey involves phases people may go through during the change process. Different people will react in different ways and go through different phases. People are likely to undergo these phases, or something similar, even if change is managed well.

The difference is if change is managed properly staff will go through these phases quickly and resistance will be minimal, and they may even avoid some of the more negative phases. However, if change is not properly managed people may get stuck in the more unproductive phases and may never accept the new order. Quiet resistance then becomes a way of life.

Another essential we must deal with is one of the less comfortable aspects of change, the probability of making people redundant. Sometimes it is unavoidable and as well as the distress for those who are made redundant is the upset and unsettlement for those who remain.

For those who have to be made redundant some will deal with the experience better than others but all will need support. In large organisations there are often very comprehensive programmes for getting people into new jobs in other organisations.

Even for small organisations, helping people with CVs, practising interviews and helping them set up a network of contacts is manageable. This helps those who have to go but also signals to those who remain that managers care and will do what they can to help.

3.2.6 Internal segments

When embarking on a programme of change employees should be segmented according to their attitude and likely response. For example:

Initiators	Those who have identified the need for change and are driving and/or encouraging change to happen.
Opponents	People who see the change as threatening and will resist. They either cannot see any justification for change or are simply not convinced. They may just want things to remain as they are.
Campaigners	People who can see the benefits or gain in some way and are therefore enthusiastic about the change. They will be keen to help drive the change through.
Neutrals	People who are yet to be convinced one way or the other. They do not feel strongly about the change and have not made up their mind whether the change is for the better or not.

This is useful for understanding internal segments. Marketers, in particular, should be able to identify the needs and motives of campaigners, neutrals and opponents.

If we understand the change journey and the very different perspectives we can better understand how we need to segment internal markets. For opponents, the journey can be a roller coaster of emotions and will appear more threatening and uncertain than for campaigners.

The opponents' perception of change tends to be negative and viewed as 'problems and trouble':

Aware	Acknowledges information.
Uncertainty	Apprehension is experienced by opponents as they see threats emerging and try to make sense of what is happening and interpreting messages.
	They will look for reassurance and hope that the *status quo* can be maintained.
Reaction	Rejection is the response of the opponent as change plans begin to be implemented and hope of maintaining the *status quo* diminishes.
	They withdraw and refuse to acknowledge what is happening.
Despair	By this stage opponents may completely withdraw or may actively resist.
	Whether or not opponents move on from this point depends on either how events are being managed and/or personal opportunities and threats.
Exploration	Beginning to come to terms with change, redefining new roles and tasks. Confidence and co-operation is growing and they begin to surface from disengagement.
Engagement	Settling down to new tasks and roles. For opponents, in particular, and for change to take hold and be a success a period of consolidation is essential.
	It is a time for reflection and learning.

The journey for campaigners is not all plain sailing. Their enthusiasm and belief in the change reduces perceptions of threat and they are happy to challenge the *status quo*. However, during some periods of the journey as the full implications are realised there will be periods of uncertainty and doubt.

The journey for campaigners usually causes less emotional turmoil. The campaigners' perception of change tends to be positive and they see possibilities and progression.

Aware	Acknowledges information.
Uncertainty	Hesitation is momentarily experienced by campaigners as they interpret information on change and work out how they are going to deal with it.
Reaction	Enthusiasm is the response of the campaigner as they see the benefits and opportunities and they embrace the change.
Despair	As campaigners begin to understand the full implications of change they may question some aspects. This is short lived as campaigners quickly move into exploring how to resolve issues.
Exploration	Campaigners are driving and redefining new roles and tasks and their continued confidence and co-operation grows.
Engagement	Settling down to new tasks and roles. Campaigners use reflection and learning as an opportunity for consolidation.

'Neutrals' will start off with no strong feelings one way or another. They will wait, listen and obverse and be swayed either by opponents or by campaigners, depending on their own susceptibility to seeing opportunities or threats and the persuasiveness of arguments from both sides. They will then start their journey, either as an opponent or campaigner.

However, some neutrals can remain just that, neither being convinced nor persuaded of the benefits of change until the end or very near the end. The issue here is that if they remain neutral they are vulnerable to becoming opponents at a time when everyone else has worked their way through the change journey and could setback progress.

If they remain neutral there is a chance they are non-participants and increasingly become disengaged as others move forward. It is impossible for them to consolidate something they are not taking part in.

The focus and effort can so easily be on the opponents, the perceived 'trouble makers ' and/or possibly the campaigners as the willing drivers of change, that those who remain neutral can be neglected or missed

altogether. However, we need to identify and manage those who are neutral and make sure they complete the journey with everyone else, achieving personal growth, a sense of achievement and consolidation.

From the above, needs and motives emerge that can help us understand the segments we have identified for our internal market and manage their concerns and responses.

Who needs convincing; holding back or guiding; who needs pulling or gently pushing? All will need support at different stages but what sort of support and when?

Reasons for resisting or supporting change are complex and varied. Some people gain more power, territory or status as a result of change so will be campaigners. Others lose power, territory or perhaps a social network and so are more likely to oppose.

In your own organisation different segments are likely to emerge and you can segment accordingly to your own specific conditions.

Activity 3

Compare and contrast the internal segments within your own organisation. What specific messages will you communicate to each of the potential segments to engage and commit them to the change management process?

3.2.7 Uncertainty (employee concerns and support)

As the implications of what we are about to do emerge, but the detail of what that means has not yet been established, people will experience uncertainty as each interpret what they do understand and know and begin to formulate the future.

Communications has a vital role in reducing perceptions of threat by informing, explaining and reassuring. Hesitation, for campaigners, often only requires information from which campaigners can learn and define the job that needs to be done. The methods can therefore be simple and straightforward, for example, information sheets and team briefings.

The apprehension of opponents requires reassurance so a very different type of communication is needed. As well as information any concerns opponents have must be addressed. Providing support through personal or group discussions give an opportunity for open debate of the issues and is a good way to deal with apprehension.

3.2.8 Change project teams

The last act of Phase 2 is to set up change project teams. We want to establish change teams that can minimise negative responses and behaviour and maximise positive.

The goal is to combine expertise and use the opportunity to design a team that will facilitate change.

It goes without saying it will not work if we try to avoid problems by putting all the campaigners of change together and all the opponents in a different group.

Quite apart from the potential imbalance of essential skills we need our campaigners championing the cause and persuading opponents of the benefits with their enthusiasm. We also need opponents to challenge and force debate and discussion around contentious issues. This helps to ensure all implications and consequences are explored.

Establishing a climate of co-operation and recognising the value of all contributions can help alleviate explosive situations or withdrawal. Managers need to pay more attention to teams and work harder at managing them during change.

3.3 Phase 3 – Scope: exploration and creativity

Analysis and evaluation does much to inform managers about the shape and design of the organisation for the future. However, this is an opportunity to explore alternatives and look for creative and innovative solutions.

This is important as changing to a marketing orientation requires a better understanding of customers and stakeholders and delivering value that is excellent and/or differentiated.

An evaluation of skills of the change teams will establish which teams/people will be better at this than others. The focus of creativity is on differentiation so ideas about systems, operations, value creation, people management etc, should be to identify new ways of doing things and new added-value possibilities that will lead to stronger competitive advantage and opportunities to differentiate.

3.4 Phase 4 – Shape: design and planning

When we have established the current conditions and have a clear understanding of our orientation and approach to customers and the gaps that exist, we can and begin to design change. To avoid failure we require an integrated and thorough approach that looks at the entire organisation and indirect implications as well as direct.

We want our initiative to be the result of joined-up thinking, design and action.

The design stage is as much about learning as it is about creating a new order.

Sometimes it is difficult to anticipate outcomes until something tangible emerges from the design and only then can sensible judgements be made about the suitability of design. Therefore, it is important to be flexible, willing to learn and if necessary change the design.

3.4.1 Design and integration of vision and values

The more clearly and powerfully you can develop and communicate a powerful change 'vision' the more successful the change will be. Without a unifying vision that the organisation's stakeholders understand and are committed to; the range of initiatives that are being developed to bring about change can be misinterpreted, misunderstood and will appear to be unrelated.

As we saw in Chapter 1 a good vision, and a good vision statement:

- Encapsulates what the organisation is trying to achieve
- Provides a rationale for the changes to be undertaken
- Provides a picture for what the organisation will look like
- Incorporates something about what is expected of the employees
- Needs to appeal to the majority of stakeholders
- Comprises realistic and attainable goals

It is, therefore, a critical part of the change process and must be communicated in words, numbers and pictures if it is to really have the impact that is required. In support it is critical that leaders and managers exhibit behaviours that are aligned to the vision – they must be seen to lead by example and 'walk the talk'.

There should be goals that determine the desired cultural values. The intention is to design cultural values into what people do and how they do it. This requires focusing on operations, people, process and task. Success will only be possible if all employees are involved.

These values will be reflected in:

Design of structure and physical work environment

- Organised around customers instead of products
- Centralised v decentralised control
- Office space and decor

- Equipment and facilities to do the job
- Additional facilities such as rest rooms, car parking

Design processes, systems and procedures

- Marketing information and communications
- Efficient and effective from customers' as well as organisation perspective
- Ease of use for people delivering value or stakeholders using process
- Health and safety

Leadership and management practices

- Professional expertise and practice
- The way work is designed and organised
- The rewards and recognition
- The authority and involvement of employees not just managers
- The type of people recruited
- The approach to learning and development

Employee preparation

- The greater part of employee preparation is the identification of learning, training and development needs. Communications also plays a role.

- The design of learning solutions including transfers and secondments. This should all be completed before moving on to Phase 5.

- Training precedes implementation and moving into new roles and taking on new tasks. In some cases this may take place on the job but, if so, experienced coaches or mentors need to be identified before implementation and available during that time.

- There may be aspects of personal development that are more likely to evolve with implementation. If so this also needs a structure and planning.

The change approach selected affects the way a vision for the change process is developed:

- If the change approach is top-down and directive the vision statement will be developed by one person such as the CEO, or the Board of Directors.

- If the change approach is more bottom-up and collaborative then a wider range of stakeholders will be involved in the vision development process

Creativity and problem-solving plays a central role throughout design and planning. Having invited employees to become involved during Phase 2 and 3, in Phase 4 employee communications become a central feature of designing and planning change.

There are two aspects to employee communications.

1 Employee consultations   we consult with them on aspects of design and change
2 Employee negotiations – we negotiate for agreement on change

3.4.2 Managing employee consultations and negotiations: inclusion v exclusion

The following guidelines are useful for managing change or other situations where consultations and negotiations are required. In changing to a marketing orientation, marketing and other managers, need to work in consultation with employees to achieve the goal.

One of the most difficult jobs a manager can become involved in is employee negotiations. Unlike individual interviews such as grievance or discipline, employee negotiations are usually with a group of people, for example between management and unions over changes in the workplace. Whatever the reason, negotiations can often result in emotionally charged situations that can quickly spiral out of control if not handled properly.

Employee consultations require skilful communications and careful preparation and follow through.

3.4.3 Preparation

Outcomes

- Focus on and clarify what you want to achieve.

- This should not be a 'I want my way ' goal as it is short-sighted and usually requires 'no matter what', resulting in unhappy, unco-operative employees.

- Being clear about the outcome allows you to explore the different options open to you.

Issues

- Examine the issues that can help bring into focus the main points to be discussed and dependent factors that are likely to arise.

- This examination requires two points of view, your own and the employees involved. Trying to see things as others see them can help clarify the problems and concerns.

Information

- Any good negotiator gathers all the facts, not just what has or will happen but also the implications and who will be involved both directly and indirectly.

- Investigate the various options open and identify common ground.

Initial possible solutions

- It is worth considering possible solutions and compromises that might have to be made.

- Begin to think this through before negotiations start when you can think clearly, identify alternative options and assess the impact of compromise.

- Be prepared for ideas on possible solutions to change during the negotiations in response to the exchange between interested parties.

3.4.4 Approach

Effective negotiators know the secret to success is persuasion: take the initiative by engaging the people you need to consult with before any formal consultation meeting.

Persuasion takes time and is often more effective in small groups or one-to-one. Look for opportunities to talk through issues, present consequences of courses of action and so on.

Marketers know the importance of managing expectations and the same is true in consultations. Presenting your case and listening to the other side in a less formal atmosphere can be very productive.

3.4.5 Consultation

Build rapport

- Even if the consultations have arisen due to conflict it is important to build rapport with all parties.

- Hopefully, some of this will have taken place during pre-negotiations.

- Respect and trust are vital if negotiations are to have successful outcomes. Rand trust are not built through hostile behaviour.

Positions

- It is important not to waste time arguing over positions and falling into the trap of defending a position, no matter how irrelevant or unrealistic.

- The danger is you become entrenched in a position from which you cannot move.

- Focus on the issues and concerns and look for areas of mutual interest.

Question and listen

- You cannot understand the other point of view without asking questions, often probing questions, that reveal all the concerns and facts.

- Questions indicate you are interested, particularly if you listen to the answers.

- They also establish exactly what is being asked for.

Exploring the options

- You are unlikely to find yourself in a situation where there are only two answers – yours and theirs.

- There are usually alternative solutions and these need to be discussed and evaluated for their suitability to both sides.

- Exploring options avoids getting stuck in one place and keeps the negotiations moving forward towards a resolution.

Timing

- Rather as a comedian's success depends on their timing so does a negotiator's. Pace the negotiations and make sure you are not presenting alternative solutions before people are ready or persuaded of the benefits.

Personalities

- Avoid negotiations becoming 'hostilities' because of the personalities involved. Focusing on emotions risks issues being sidetracked.

- Try to understand things from different points of view, use active listening skills and avoid barriers to listening, use simple language and avoid being misinterpreted through ill concealed resentment.

Agreeing solutions

- Before entering into negotiations it is important to establish authority for solutions.

- Does it rest with you or will you have to get approval from a higher authority? Conclude the negotiations, summarising the concerns and issues and the agreed solutions.

3.4.6 Treaty

A written report of the consultations with agreed solutions must be circulated as soon as possible. Allow time for response and then ensure any agreed actions are implemented. These actions may require support in various forms, for example, training or counselling and arrangements must be made for any support needed.

Reaction (employee concerns and support)

As people come to terms with the implications of change they will form their own opinions on whether or not they support the change. Rejection may be unavoidable for some people.

You are dealing with emotions and feelings so sensitivity and tact is vital. Provide support for those rejecting by allowing them to articulate their concerns and help them translate threats into opportunities. For the enthusiasts allow freedom to examine and define new roles and guide and support where required.

Despair or doubt (employee concerns and support)

Despair for opponents or doubt for campaigners may begin to set in at this stage or during implementation. This is a low point when they will or may become disengaged. Underlying concerns will begin to emerge and people may suffer feelings of low self-esteem and confidence. Their doubts or despair may result in them becoming less co-operative.

Provide support by encouraging staff to explore the positive aspects of change, opportunities that exist and get threats into perspective. Help them see how their strengths fit and the contribution they can make.

Reassure them that this is a journey of discovery for all and that others will have similar feelings. We are all learning during this process. Individual discussions are likely to feature more often in communications.

Global case study

John Kotter's highly regarded books *Leading Change* (1995) and the follow-up *The Heart Of Change* (2002) describe an eight-step model for understanding and managing change. Each stage acknowledges a key principle, as identified by Kotter, relating to people's response and approach to change, in which people see, feel and then change.

Kotter writes, *"there are still more mistakes that people make, but these eight are the big ones. In reality, even successful change efforts are messy and full of surprises"*.

Kotter's eight-step change model can be summarised as:

1 **Increase urgency** – inspire people to move, make objectives real and relevant.

2 **Build the guiding team** – get the right people in place with the right emotional commitment, and the right mix of skills and levels.

3 **Get the vision right** – get the team to establish a simple vision and strategy, focus on emotional and creative aspects necessary to drive service and efficiency.

4 **Communicate for buy-in** – Involve as many people as possible, communicate the essentials, simply, to appeal and respond to people's needs. De-clutter communications – make technology work for you rather than against.

5 **Empower action** – Remove obstacles, enable constructive feedback with lots of support from leaders – reward and recognise progress and achievements.

6 **Create short-term wins** – Set aims that are easy to achieve – in bite-size chunks. Manageable numbers of initiatives. Finish current stages before starting new ones.

7 **Don't let up** – Foster and encourage determination and persistence – ongoing change – encourage ongoing progress reporting – highlight achieved and future milestones.

8 **Make change stick** – Reinforce the value of successful change via recruitment, promotion, new change leaders. Weave change into culture.

http://www.businessballs.com/changemanagement.htm

Accessed 25 March 2010

Activity 4

What are the key factors which will assist with an ineffective management of change within your own organisation?

4 Preparing a plan for change

4.1 Phase 5 – Action: implementation and control

Once agreement has been reached on all aspects of change and design completed, plans for implementing change begins. There will still be the need for consultations and negotiations during this stage but the focus of communications changes to updates on progress.

Setting new objectives signals a marketing orientation and objectives can include customer satisfaction and loyalty, employee performance (particularly customer service) and morale.

4.1.1 Implementation and control

It is now time to transfer strategy into tactical action plans, a new way of doing things.

Each section, department and team incorporates the new values as an integral part of their work:

- The way in which they co-operate and collaborate
- Taking a pride in their work
- Meeting quality targets
- Meeting or exceeding targets
- Taking initiatives to improve and solve problems

Action (or project) plans determine who is doing what, when, where and how. Objectives, budgets, schedules and measures are all part of the action plans.

Personal targets and performance standards are agreed with individuals and teams.

Authority is delegated with responsibility. Pay attention to the scheduling of events.

Control of the change process requires measurements that link back to the objectives, setting-up feedback mechanisms for monitoring and evaluating progress and appraisals, possibly more regular than usual. The feedback mechanisms and appraisals allow managers and employees to learn during the change and make any necessary adjustments and modifications. Creativity and problem-solving skills will continue to be needed at this stage.

During implementation continue to communicate and keep the momentum going. For example, you can use visual effects to show how things are improving and the progression towards success, including:

- Moving graph of reduced complaints
- Moving graph of improving organisation image
- Changing from current position to desired position
- Improving sales/profits

This is linked back to team and individual performance to be meaningful.

Exploration (employee concerns and support)

- Allow exploration, experimentation and focus on priorities.

- Employees need to feel in control of events and that they are genuinely being given responsibility for shaping the changes and improvements.

- Provide support by giving feedback, guiding and coaching or mentoring where required.

Engagement (employee concerns and support)

- Some employees will become engaged earlier than others and for some probably during implementation.

- Others may not be fully engaged until there is time to consolidate and assess where they have arrived.

- Longer-term goals can now be established.

- Provide support through recognition and reward for progress and success.

- Controls and supervision change at this stage from controlling the change to the usual controls on performance.

- Employees can take more responsibility for moving forward.

4.1.2 Managing the transition at the organisational level

Earlier in the chapter we referred to Lewin's Force Field Analysis as a model by which leaders and managers can establish the restraining forces that may hinder change.

The model also identifies three states that must be managed carefully:

1   **Unfreezing**

  - Making people within the organisation ready for change by making them aware of the need for the change.

  - Creating a 'readiness for change' among all stakeholders at all levels.

2   **Moving**

  Implementation of the needed changes through the selected approaches and activities.

3   **Refreezing**

  - Embedding changes throughout the organisation to ensure stakeholders do not relapse into old patterns of behaviour.

  - Given that many organisations in today's global business environment are often in an ongoing state of adaptation it is becoming more common to refer to the unfreeze phase as mobilisation and the refreeze phase as "sustain" or "institutionalise".

4.1.3 Transition at the individual level

Adams, Hayes and Hopkins (1976) developed the transition curve to describe seven phases that individuals go through during change.

STAGE	INDIVIDUAL FEATURES
1 Shock	Encounter the need for change. Self-confidence takes a knock as they recognise they need to do things differently.
2 Denial	Try to rationalise change and convince themselves that they will not need to change. Self-confidence increases.
3 Awareness	Recognition of need for personal change brings a drop in confidence as they become aware of their inadequacy to fulfil new role.
4 Acceptance	Can move forward when recognise need to break from the past, let go of old attitudes and behaviours and adopt new ones.
5 Testing	Identifying and testing new behaviours, maybe through training and professional development. New behaviours build confidence.
6 Search	Assimilate learning from their successes and failures, begins to understand why some behaviours work and others fail.
7 Integration	Integrating new behaviours into day-to-day work.

Based upon research undertaken by Adams *et al* (1976)

4.1.4 Linking individual and organisational transitions

Each phase of organisational transition can be linked to individual change with appropriate measures being put in place:

Organisational phase	Measures	Barriers
Mobilise	• Help individual let go of past • Minimise shock • Communicate intentions as early as possible • Expect resistance	• Disinformation • Lack of information • Isolation • Only seeing the negative
Move	• Help individuals change through coaching, support, giving and receiving feedback • Putting in place education, training, personnel development, new working practices and systems	• Opinions not allowed • Uncertainty • Lack of forums for expression • Punishing failure • Resistance • Keep celebrating the past
Sustain	• Support individuals in new roles • Encourage reflection on change and learning • Celebrate success • Reinforce new behaviours	• Moving on to next change initiative • No goals or targets

In developing the transition process there are a number of questions to consider in developing an outline and the sequence in which the change levers and interventions should be deployed:

- Is there a coherent strategy that is understood and shared throughout the organisation?
- Are supporting systems and structures under development?
- Is there a trigger or change or has one been manufactured?
- Are there visible 'early wins' designed into the change process?
- Are day-to-day activities aligned to get required outputs?
- Are the identified barriers to change being removed or dealt with?
- Are changes supported with symbolic activity?
- Is communication built into the change process?

4.1.5 Symbolic breaks with the past

Doing things differently can indicate a break from the past can include for example:

- Change of dress
- Moving to an open plan office environment
- Ensuring senior management are more accessible through "open-door" policy
- Change of language
- Change promotion policies
- Moving people into different offices
- Organisational restructures
- Removing old furniture, paintings symbols of the past etc

4.1.6 Monitoring progress

There can be a number of unintended consequences of change management deliberate or otherwise:

- **Ritualisation of culture change**
  Culture change becomes a once-a-year ritual

- **Hijacked process**
  Others hijack process as a means to promote their own projects

- **Cultural erosion**
  Other goals and objectives become more a priority than those relating to culture change

- **Cultural reinvention**
  Superficial change which camouflages the continuation of old culture

- **Ivory tower change**
  Change cannot be implemented in a meaningful way

- **Inattention to symbolism**
  Old symbols, attitudes and routines are left in place

- **Behavioral compliance**
  Employees follow new rules and procedures without changing their beliefs

It is necessary to understand how people interpret things and why; which events and activities are having an impact; how they are being interpreted and how it is affecting behaviour and performance.

To avoid senior management being the "last to know" monitoring mechanisms can include:

- Focus groups and workshops
- Management by wandering around
- Team briefings
- Question and Answer sessions
- Independent research undertaken by external consultants
- Staff representative collating feedback
- Staff suggestion schemes
- Confidential "hotlines"
- Attitude survey and questionnaires

Global case study

When Bass plc announced in 200 it was putting its brewing operation up for sale it resulted in two years of uncertainty for its employees before it was purchased by Coors in 2002. During this period of uncertainty a number of mechanisms were deployed to keep staff informed and maintain open communication. Team briefings were used to generate two-way dialogue between management and their teams, a Q&A provided instant feedback through the intranet, focus groups dug deeper into issues and surveys were sent to a random sample every few months to check progress and quality of communications from the staff's perspective.

Adapted from *Exploring Corporate Change*, Balogun, J. and Hailey, H., (2004) (2nd edition). FT Prentice Hall.

4.2 Phase 6 Unite: consolidation and learning

Even if you work in an organisation where continuous change and improvement is a way of life managers have a duty to provide a sense of achievement and a period of consolidation. Allow time for employees to become familiar with new work routines and practices.

There may still be some training, coaching or mentoring required. Employees will either be re-enforcing old social networks or building new and these networks should be encouraging behaviour that reflects the desired values of a marketing orientation.

This is a time to embed new ways of working. People are still adapting to a different way of doing things and exploration of what works, what needs adjusting may still be required.

Provide opportunities for review and learning from the process of change, the experiences and activities. Confront the scary things that happened, share the doubts and concerns in an open and supportive way.

This learning is not just a therapeutic session, but is to inform any further change and improvement initiatives so that problems and mistakes can be minimised or avoided and successes can be repeated.

Global case study

Staples, the global retail stationers, are an example of an organisation that recognises the direct link between leadership and high organisation performance. Their HR Director spends a lot of time thinking and talking to leaders about how to achieve high performance. He does not believe it is an HR issue but rather a leadership challenge and the challenge is how to negotiate a balance between the organisation (its structure, culture, systems and leadership style) and its people (attitudes, behaviour, skills and output). The HR Director argues that too often, in response to changing trends and the increasing sophistication and demands of customers, leaders focus on changing structures and processes rather than attitudes and behaviour, where a real valuable difference can be achieved. For leaders who focus on changing operations the usual result is people being able to take less initiative. Leaders need to focus how they get people to take responsibility for change that can make a positive and relevant difference. The leadership challenge is how to create the right environment where people are motivated and willing to take responsibility for developing appropriate attitudes and behaviour and given freedom to innovate. How do leaders move from managing this initiative to leading?

Staples are in the process of creating such an environment. They have started by helping leaders across the organisation understand more about the results they have been getting, and how they are directly linked to beliefs and assumptions they hold about themselves or the organisation. This new understanding helps them to make positive choices about work and their lives. The results have been excellent with an observable change in behaviour and Staples can see managers developing into leaders who motivate.

1 Evaluate the external and internal drivers of change and barriers to change

- The need for change can be triggered by either internal or external triggers.

- External changes occur in seven broad areas all of which are linked either tentatively or strongly.

- We can use the McKinsey Seven S Model (Waterman et al, 1980) to identify what might trigger change internally. As with external drivers, internal drivers are linked and a change in one area affects what is happening in other areas.

- Leaders need to understand these connections.

- Reactions to change are varied and can be positive particularly if the organisation is clearly in difficulty.

- The skill of leaders is to recognise when it is appropriate to change the organisation in response to external changes and when it is not.

- When the scope of improvements is larger and widespread this will lead to more momentous change.

2 Critically assess methods of change and its impact on the marketing strategy

- The implementation of a marketing strategy will attempt to bring about strategic change. Where this occurs it may be referred to as strategic renewal.

- This process might consist of a few large steps or numerous small ones. Strategic change measures can be evenly spread out over an extended period, allowing the organisation to follow a relatively steady pace of strategic renewal.

- The pace of organisational change can be split into two related parts: timing of change and the speed of change.

- Organisations need to decide whether to achieve revolutionary (disruptive) or evolutionary (gradual) change.

- Once leaders choose to adopt revolutionary or evolutionary change they will find it very difficult to simultaneously or subsequently use the other.

3 Rationalise a process for managing change

- Change management is the deliberate strategy to minimise the negative impact of change.

- The challenge for the leader is to pull together the process of managing change with managing people through change.

- Where change may impact on stakeholders leaders will need to develop a strategy and process for managing those relationships.

- Leaders of change need to work on creating an atmosphere of honesty and openness.

- Leaders are often the initiators of change (but not exclusively) so as to pursue their vision and future strategic direction and are rarely interested in the detail of implementation.

- We need to understand the effect change will have on people, different people who will react in different ways and undergo different phases.

- If change is managed properly they will go through these phases quickly, resistance will be minimal, and they may even avoid some of the more negative phases.

- When embarking on a programme of change employees should be segmented according to their attitude and likely response. Success will only be possible if all employees are involved.

- Creativity and problem-solving will play a central role throughout the design and planning of a change programme which will go through several stages of management. Each stage needs to acknowledge people's response and approach to change

4 Evaluate how to best prepare a plan for change

- Once agreement has been reached on all aspects of change, and design completed, plans for implementing change can begin.

- Action (or project) plans need to be drawn up to determine who is doing what, when, where and how.

- Control of the change process requires measurements that link back to the objectives and the setting-up of feedback mechanisms for monitoring and evaluating progress.

Activity 1

This will very much depend on your organisation but the sort of things you will be looking for are, for example, technology, operational or system change that leads to increased efficiency; customer value change that leads to increased differentiation and customer demand and loyalty. Was it obvious to all why the change was taking place and the consequences of not changing? Were people involved and motivated? Did the change achieve the objectives?

Activity 2

This will very much depend on your organisation. Be aware of the differences between the two approaches and the different styles that the leader will need to take. It will also depend to a large extent on the prevailing culture within the organisation and its readiness for change. Also consider the external environmental forces at work and the pace of change occurring.

Activity 3

This will very much depend on your organisation. Internal markets need to be understood if resistance to change and perceptions of threats are to be minimised.

Change can only be successful if the people responsible for managing and implementing change accept the change. To understand these internal markets we need to understand how people perceive the intended changes and what affect it will have on them. Different people will be affected in different ways and, if we know how they are affected, we can set up support systems and training programmes that will help them come to terms with the change and take on their new tasks and responsibilities. Internal markets will typically be segmented as drivers of change, resistors and neutrals.

Activity 4

This will very much depend on your organisation and its approach but typically it will include:

- Clear objective of the purpose of change
- Planning for change
- Communications – internal marketing
- Segmenting internal market
- Involving people – project and design teams
- Learning and development
- Support systems
- Evaluation

Adam, J., Hayes, J. and Hopkins, B., (1976). *Transition Understanding and Managing Personal Change*. London: Martin Robertson & Co.

Balogun, J. and Hailey, H., (2004). *Exploring Corporate Change*. London: FT Prentice Hall.

De Wit, B. and Meyer R., (2005). *Strategy Synthesis: Resolving Strategy Paradoxes to Create Competitive Advantage*. London: Thomson.

Kotter, J., (1995). *Leading Change*. Boston: Harvard Business School Press.

Kotter, J. and Cohen, D.S., (2002). *The Heart of Change*. Boston: Harvard Business School Press.

Lasserre, P. and Schutte, H., (2006). *Strategies for Asia Pacific: Meeting New Challenges,* 3rd edition. Basingstoke: Palgrave Macmillan.

Lewin, K., (1951). *Field Theory in Social Science: Selected Theoretical Papers*. New York: Harper.

Waterman, R. H., Peters, T. J. and Phillips, J. R., (1980). 'Structure is not organisation'. *McKinsey Quarterly in-house journal*. McKinsey & Co.: New York.

Index